Radon Series on Computational and Applied Mathematics 7

Radon Series on Computational and Applied Mathematics

1 *Lectures on Advanced Computational Methods in Mechanics*
Johannes Kraus and Ulrich Langer (eds.), 2007

2 *Gröbner Bases in Symbolic Analysis*
Markus Rosenkranz and Dongming Wang (eds.), 2007

3 *Gröbner Bases in Control Theory and Signal Processing*
Hyungju Park and Georg Regensburger (eds.), 2007

4 *A Posteriori Estimates for Partial Differential Equations*
Sergey Repin, 2008

5 *Robust Algebraic Multilevel Methods and Algorithms*
Johannes Kraus and Svetozar Margenov, 2009

6 *Iterative Regularization Methods for Nonlinear Ill-Posed Problems*
Barbara Kaltenbacher, Andreas Neubauer and Otmar Scherzer, 2008

Jan H. Maruhn

Robust Static Super-Replication of Barrier Options

Walter de Gruyter · Berlin · New York

Keywords
Static hedging, barrier options, robust optimization, stochastic volatility, semi-infinite optimization, semidefinite programming.

Mathematics Subject Classification 2000
49-02, 91-02, 91B28, 90C30, 90C34, 91B24.

♾ Printed on acid-free paper which falls within the guidelines of the ANSI to ensure permanence and durability.

ISBN 978-3-11-020468-1

Bibliographic information published by the Deutsche Nationalbibliothek

The Deutsche Nationalbibliothek lists this publication in the Deutsche Nationalbibliografie; detailed bibliographic data are available in the Internet at http://dnb.d-nb.de.

Printed in Germany
Cover design: Martin Zech, Bremen.
Printing and binding: Hubert & Co. GmbH & Co. KG, Göttingen.

To Sylvia

Preface

Since the ground-breaking work of Black and Scholes it is common practice to hedge exotic options by dynamically adjusting portfolio positions based on the sensitivities (Greeks) of the target option. However, while this works sufficiently well in practice for instruments with well-behaved sensitivities, it may be impossible to dynamically hedge options with discontinuous payoff profile and hence wildly moving Greeks. Motivated by this problem as well as the idea to avoid a continuous adjustment of the portfolio positions, the concept of static trading strategies has been developed in the literature.

Due to the fact that barrier options are components of a variety of products traded in the market, various authors have focussed on the static hedging of barrier options with discontinuous payoff profile like up-and-out calls. For these options, a static trading strategy consists of buying a portfolio of standard calls and holding this portfolio constant until either the barrier option expires or the barrier is hit. In the latter case of a barrier hit at some time in the future, the portfolio of calls is liquidated and hopefully provides a payoff equal to the value of the up-and-out call.

As the future value of the calls changes with the volatility surface, it is intuitively clear that a static hedging strategy has to be robust against changes of this surface. However, this key property has so far not been throroughly adressed in the literature, because considering the dynamics of the volatility surface leads to analytical and computational difficulties.

The book at hand closes this gap by developing an optimization approach which combines the idea of static hedging with the concept of super-replication and finally adds robustness against movements of the volatility surface in the sense of a worst case design. The resulting robust static hedging problem is analyzed in detail from a theoretical and numerical point of view. This analysis draws from the fields of financial mathematics, stochastic and semi-infinite optimization, convex analysis, partial differential equations as well as semidefinite optimization, to derive appropriate existence, duality and convergence results.

Detailed examples show that the proposed static hedging framework is superior to the approaches developed in the literature. In particular the computed robust static hedging strategies prevent potentially huge losses due to changes of the volatility surface and only consist of a small number of liquidly traded standard options. Surprisingly, the robustness can be gained by relatively little additional cost. Moreover, in contrast to other approaches the computations prove to be numerically stable due to an implicit regularization of the hedging problem.

The book develops the static super-replication approach in several steps. Chapter 1 summarizes the theoretical tools needed throughout the book, including results from the fields of mathematical finance and optimization. Afterwards we introduce the concept of statically hedging barrier options in Chapter 2 and review the main approaches that have been considered in the literature so far.

Chapter 3 presents the starting point of our analysis in the form of a stochastic optimization problem characterizing cost-optimal static super-replication strategies in general financial market models. As it turns out, the structure of the problem allows to prove the existence of optimal static hedging strategies based on suitable results from the field of convex analysis. In addition we introduce a first naive Monte Carlo discretization of the stochastic super-replication constraint which allows to compute the strategy by solving a large scale linear programming problem. Numerical results for the Black–Scholes and Heston's stochastic volatility model show that the proposed approach outperforms all other approaches in the literature. In particular the computed portfolio matches the sensitivities of the target option and is only marginally more expensive than the barrier option itself.

Although these results are promising, the Monte Carlo method has the significant drawback of slow convergence which can lead to long computation times. To circumvent this problem, Chapter 4 reformulates the stochastic super-replication constraint into an infinite number of deterministic constraints. The structure of the resulting linear semi-infinite optimization problem can be exploited to drastically reduce computation time in comparison to the Monte Carlo-based method. Moreover, the equivalent formulation allows to derive the dual optimization problem with an interesting economical interpretation. In analogy to the case of dynamic hedging the dual problem maximizes the discounted expectation of the barrier option payoff. But this time the expectation is maximized over the set of measures which are price-consistent with respect to the standard options in the hedge portfolio – a far bigger set than the equivalent martingale measures. This also theoretically underlines that the ability to approximate call prices is a key property for the derivation of a successful static hedge.

Based on the semi-infinite reformulation of the hedging problem Chapter 5 introduces a robustification of the hedging strategy against changes of the volatility surface. From an optimization point of view, the robustness is described by model parameter uncertainty sets which correspond to an infinite number of future volatility surface scenarios. The existence of solutions of this problem is proven by applying the maximum principle and Mizohata's uniqueness theorem for parabolic partial differential equations. Furthermore, we prove the convergence of suitable algorithms for the numerical solution of the robust hedging problem. These methods for the first time quantify and eliminate the model parameter sensitivity of hedge portfolios for barrier options by addressing the nonlinearity of call option prices with respect to changes of the volatility surface. Numerical results for the Black–Scholes and Heston's stochastic volatility model show that robustness against model parameter uncertainty can be gained by surprisingly low cost.

Chapter 6 shows that the previous results can in analogy be transfered to a sub-replication setting. Combined with the super-replication prices this leads to robust static bounds for the price of barrier options which are surprisingly tight if compared to purely model-independent bounds in the literature. Moreover, we also robustify the static hedging problem against jumps by introducing an additional model parameter uncertainty set. However, based on the maximum principle, we can eliminate this uncertainty set and transform the problem to a formulation which can be interpreted as moving the barrier. The chapter concludes with a section which shows how to apply the robust hedging approach to a large variety of barrier option contracts.

Chapter 7 considers an additional robustification against model errors in the form of ellipsoidal uncertainty sets in the price space. Fortunately, these ellipsoidal uncertainty sets can be replaced by an infinite number of second order cone or semidefinite programming constraints. To compute a solution of the problem, we propose an algorithm successively solving linear and nonlinear semidefinite programming problems and prove its convergence under mild conditions.

Finally, Chapter 8 analyzes the real world performance of the static superhedge based on a seven year dataset. As it turns out, the theoretical robustness against changes of the volatility surface (including the skew) is also confirmed empirically. Compared to a static strike spread hedge and a dynamic local volatility hedge, the robust static superhedge is the portfolio leading to the smallest hedge error dispersion.

To summarize, the book develops a static hedging approach which allows to robustify trading strategies against model parameter uncertainty in the form of volatility shocks, skew and jump risk as well as model errors. The practical applicability of the framework as well as the performance of the trading strategies are illustrated with numerous examples throughout the book.

Some Words of Gratitude

This book is based on the author's dissertation accepted on March 5th, 2007, at the University of Trier. Most of the work evolved during my time as a research associate at the Department of Mathematics, in the research group Numerical Analysis led by Prof. Dr. Ekkehard W. Sachs. At this point I would like to thank the people who have contributed to this research and supported me during the past years.

First of all I would like to thank my advisor Prof. Dr. Ekkehard W. Sachs for numerous discussions as well as the joint work and publications on static hedging. Besides his contributions to this work, I am especially grateful that he encouraged me to experience all aspects of the life of a researcher including the freedom to choose my academic focus, the opportunity to participate in various international conferences and that he always supported my ambitions to work on projects with external industry partners. It was a very inspiring and pleasant time in the research group.

Furthermore, I wish to thank Dr. Hansjörg Albrecher for fruitful comments and for acting as a referee of my PhD thesis.

In addition I would like to thank the Financial Engineering team (Equities, Commodities and Funds) of UniCredit Markets & Investment Banking, Bayerische Hypo-

und Vereinsbank AG, for supporting and initiating my work on static hedging during a joint research project. I am particularly indebted to Alexander Giese for very deep and enduring discussions on this work and the practice of mathematical finance in general. Our initial paper on static hedging was the starting point for the research presented in this book.

Moreover, I wish to express my gratitude to Dr. Friedemann Leibfritz for detailed discussions on the theory and practice of optimization as well as the joint work on the robustification against model errors. My thanks also go to Morten Nalholm and Matthias Fengler for the collaboration regarding the empirical performance of the robust static hedge. I also thank my colleagues Ewgenij Hübner, Ilia Gherman and Christina Jager at the University of Trier for a great time in the department of mathematics.

Special thanks go to PD Dr. Robert Plato for numerous suggestions and for being so patient with respect to necessary revisions of the manuscript. Moreover, I am grateful for the detailed comments and suggestions of an anonymous referee.

Finally, I would like to thank my wife Sylvia for constantly supporting me over the past years. Without her the completion of this research would not have been possible. I am also indebted to my parents for their moral and financial support of my academic career.

Jan Maruhn
Munich, April 2009

Contents

1 Theoretical Background

In this chapter we review some basic mathematical concepts which we refer to throughout the book. Section 1.1 briefly introduces some definitions from probability theory and the field of stochastic differential equations. Afterwards Section 1.2 presents the main ideas of mathematical finance which are relevant in our context. In Section 1.3 we finally discuss several classes of optimization problems with a particular focus on the definition and properties of linear semi-infinite programming problems.

1.1 Probability Theory and Stochastic Differential Equations

The concepts we introduce in this section briefly summarize the main probabilistic definitions we need during the remainder of this work. A more detailed discussion of the presented topics can be found in Bauer [7] and Shiryaev [81].

As a starting point we assume the existence of a probability space $(\Omega, \mathcal{F}, P)$, where $\mathcal{F}$ denotes a sigma-algebra over the set $\Omega \neq \emptyset$ and P is a suitable probability measure. In this context a set $N \in \mathcal{F}$ is a *null set* if and only if $P(N) = 0$. A property, defined for all $\omega \in \Omega$, is said to hold *almost surely (a.s.)* if there exists a null-set N such that the property holds for all $\omega \in N^c$. Examples in our setting include inequalities or equalities of real-valued random variables $X, Y : \Omega \to \mathbb{R}$ to hold almost surely, e.g. $X \geq Y$ (a.s.) or $X = Y$ (a.s.).

If $B \in \mathcal{F}$ is a given set, the measure P restricted to B is defined by $P|_B(\cdot) := P(\cdot \cap B)/P(B)$. Furthermore, the measure $P^{(X,Y)}$ denotes the joint distribution of two random variables X and Y under P. In case $(\Omega, \mathcal{F}) = (\mathbb{R}, \mathcal{B})$, where $\mathcal{B}$ is the Borel sigma algebra, $\text{supp}(P)$ denotes the support of P, that is the smallest closed set with measure 1.

For $I \subset \mathbb{R}_+$ the family $\mathbb{F} = (\mathcal{F}_t)_{t \in I}$ is called a *filtration* on the measurable space $(\Omega, \mathcal{F})$ if $\mathcal{F}_t \subset \mathcal{F}\ \forall t$ and $\mathcal{F}_s \subset \mathcal{F}_t$ for $s, t \in I$, $s < t$. If we further denote the set of all null sets by $\mathcal{N}$, i.e. $\mathcal{N} = \{N \in \mathcal{F} : P(N) = 0\}$, then the *augmented filtration* $\mathbb{F}^* = (\mathcal{F}_t^*)_{t \in I}$ of $\mathbb{F} = (\mathcal{F}_t)_{t \in I}$ is the family of smallest sigma-algebras $\mathcal{F}_t^*$ containing $\mathcal{F}_t \cup \mathcal{N}$.

Given a filtration $\mathbb{F} = (\mathcal{F}_t)_{t \in I}$, a random variable $\tau : \Omega \to I \cup \{\sup I\}$ is called $\mathbb{F}$-*stopping time* if $\{\omega \in \Omega : \tau(\omega) \leq t\} \in \mathcal{F}_t$ for all $t \in I$.

A family of random variables $X = (X_t)_{t \in I}$, $X_t : \Omega \to \mathbb{R}^m$ defines a *stochastic process* on $(\Omega, \mathcal{F}, P)$. For fixed $\omega \in \Omega$ the mapping $t \mapsto X_t(\omega)$ describes a *path* of the process $(X_t)_{t \in I}$. The process X is said to be *adapted to the filtration* $\mathbb{F}$ if X_t is $\mathcal{F}_t$-measurable for all t. X is $\mathbb{F}$-*progressively measurable* if $I = [0, T]$ and the mapping $[0, t] \times \Omega \to \mathbb{R}$, $(s, \omega) \mapsto X_s(\omega)$ is measurable with respect to the product sigma-algebra $\mathcal{B} \otimes \mathcal{F}_t\ \forall t \geq 0$. Furthermore, we define the filtration $\mathbb{F}^X = (\mathcal{F}_t^X)_{t \in I}$

generated by X as the family of smallest sigma-algebras $\mathcal{F}_t^X$ such that X_s is $\mathcal{F}_t^X$-measurable for all $s \leq t$. The probably most well-known example of a stochastic process is implicitly defined via the following axioms.

Definition 1.1. Let $(\Omega, \mathcal{F}, P)$ be a probability space with filtration $\mathbb{F} = (\mathcal{F}_t)_{t \in I}$, $I = [0, T]$, $0 < T < \infty$, or $I = \mathbb{R}_+$. Then a stochastic process $W = (W_t)_{t \in I}$, $W_t : \Omega \to \mathbb{R}$, is called an $\mathbb{F}$-Wiener process (or $\mathbb{F}$-Brownian motion) if

i) $W_0 = 0$ (a.s.)
ii) $W_t - W_s$ is independent from $\mathcal{F}_s$ for all $s, t \in I, 0 \leq s < t$
iii) $P^{W_t - W_s}$ is normally distributed with mean zero and variance $t - s$ for all $0 \leq s < t$ with $s, t \in I$
iv) The mapping $t \mapsto W_t(\omega)$ is continuous (a.s.).

Furthermore, a process $W = (W_t^1, \ldots, W_t^m)_{t \in I}$, $W_t^i : \Omega \to \mathbb{R}$, $i = 1, \ldots, m$, is called a multidimensional $\mathbb{F}$-Wiener process with correlation matrix $\Lambda = (\rho_{ij})_{i,j} \in \mathbb{R}^{m \times m}$ if and only if W^i is an $\mathbb{F}$-Wiener process for each i and the correlation of W_t^i and W_t^j is ρ_{ij}.

Note that for the special case of a two-dimensional $\mathbb{F}$-Brownian motion $W = (W^1, W^2)$ the correlation matrix Λ is uniquely identified by $\rho_{1,2} = \rho_{2,1} =: \rho$. Hence, in this particular situation, the two Brownian motions W^1, W^2 are said to be correlated with correlation coefficient ρ.

Besides the Brownian motions, the martingales form another class of stochastic processes which play an important role in the theory of mathematical finance.

Definition 1.2. Let $X = (X_t)_{t \in I}$, $I \subset \mathbb{R}_+$, be a real-valued stochastic process adapted to the filtration $\mathbb{F}$ with $X_t \in L^1(P)\ \forall\, t \in I$. Then X is called an $(\mathbb{F}, P)$-martingale if and only if

$$E(X_t | \mathcal{F}_s) = X_s\ (a.s.)\ \forall\, s, t \in I,\ s < t.$$

Further X is a local martingale if there exists a sequence of stopping times $(\tau_k)_{k \in \mathbb{N}}$, $\tau_k \leq \tau_{k+1}$, $\tau_k \to_{k \to \infty} \infty$ (a.s.) $\forall\, k$, such that $X^{\tau_k} = (X_{\tau_k \wedge t})_t$ is a martingale $\forall\, k$.

Here $L^1(P)$ is the space of measurable, real-valued functions whose absolute value has a finite integral with respect to P. Further $E(X_t | \mathcal{F}_s)$ denotes the conditional expectation of X_t relative to $\mathcal{F}_s$.

Without presenting the details we mention at this point that stochastic Itô-integrals are one example for local martingales. Based on this observation, the idea arises to define a stochastic process which is decomposed in a deterministic drift part and a martingale part. This idea is formalized in the following definition.

Definition 1.3. Suppose $(\Omega, \mathcal{F}, P)$ is a probability space with filtration $\mathbb{F} = (\mathcal{F}_t)_{t\geq 0}$ and $W = (W_t)_{t\geq 0}$ is an $\mathbb{F}$-Brownian motion. Then an Itô process $(X_t)_{t\geq 0}$ is a process of the form

$$X_t = X_0 + \int_0^t H_s ds + \int_0^t K_s dW_s, \tag{1.1}$$

where X_0 is $\mathcal{F}_0$ measurable, K and H are $\mathbb{F}$-progressively measurable and

$$\int_0^t |H_s| ds < \infty \ \forall\, t \ (a.s.), \quad \int_0^t K_s^2 ds < \infty \ \forall\, t \ (a.s.). \tag{1.2}$$

In (1.1) the term $\int_0^t K_s dW_s$ denotes the stochastic integral of K with respect to the Brownian motion W. The conditions (1.2) guarantee that this integral as well as $\int_0^t H_s ds$ are well defined.

Equipped with this notation, we can move on to the concept of stochastic differential equations. Instead of defining a stochastic process explicitly as we did in Definition 1.3, we now wish to define a process by its initial value and its dynamics over time.

Definition 1.4. Assume that a probability space $(\Omega, \mathcal{F}, P)$ with augmented filtration $\mathbb{F} = (\mathcal{F}_t)_{t\in I}$, $I = [0, T]$ is given. Further let $W = (W_t^1, \dots, W_t^k)_{t\in I}$ be an $\mathbb{F}$-Brownian motion and assume that $\beta = (\beta_1, \dots, \beta_m) : \mathbb{R}_+ \times \mathbb{R}^m \to \mathbb{R}^m$, $\gamma = (\gamma_{ij})_{i,j} : \mathbb{R}_+ \times \mathbb{R}^m \to \mathbb{R}^{m\times k}$ are Borel-measurable functions. Suppose that $\zeta = (\zeta_1, \dots, \zeta_m)$ is an $\mathbb{R}^m$-valued, $\mathcal{F}_0$-measurable random variable. Then an $\mathbb{R}^m$-valued process $X = (X_t^1, \dots, X_t^m)_{t\in I}$ is a solution of the system of stochastic differential equations

$$dX_t^i = \beta_i(t, X_t)dt + \sum_{j=1}^k \gamma_{ij}(t, X_t)dW_t^j, \ X_0^i = \zeta_i, \ 1 \leq i \leq m \tag{1.3}$$

with initial condition $X_0 = \zeta$, if the integrals $\int_0^t \beta_i(s, X_s)ds$, $\int_0^t \gamma_{ij}(s, X_s)dW_s^j$ are well defined for all t and

$$X_t^i = \zeta_i + \int_0^t \beta_i(s, X_s)ds + \sum_{j=1}^k \int_0^t \gamma_{ij}(s, X_s)dW_s^j \ (a.s.) \ \forall\, t \in I, \ 1 \leq i \leq m.$$

For a more detailed introduction to the theory of stochastic differential equations we refer the reader to the books of Øksendal [75] or Karatzas and Shreve [57]. In both books the authors further prove variants of the following existence result for the solution of stochastic differential equations.

Theorem 1.5. *Based on the notation and assumptions of Definition 1.4 we further suppose that $\|\zeta\|_2$ is square integrable and that the following conditions hold:*

i) Lipschitz condition: $\forall\, T > 0\; \forall\, N > 0\; \exists\, c_1(T,N) < \infty$ *such that* $\forall\, t \in [0,T]$, $\forall\, x, y \in \mathbb{R}^m$ *satisfying* $\|x\|_2, \|y\|_2 \le N$ *we have*

$$\|\beta(t,x) - \beta(t,y)\|_2 + \||\gamma(t,x) - \gamma(t,y)\||_2 \le c_1(T,N)\|x-y\|_2$$

ii) Growth-condition: $\forall\, T > 0\; \exists\, c_2(T) < \infty$ *such that* $\forall\, t \in [0,T]$, $\forall\, x \in \mathbb{R}^m$

$$\|\beta(t,x)\|_2^2 + \||\gamma(t,x)\||_2^2 \le c_2(T)(1 + \|x\|_2^2).$$

Then the stochastic differential equation (1.3) has an almost surely unique solution.

Proof. See Øksendal [75], Theorem 5.2.1 and the weakening of the Lipschitz condition in Irle [54], Theorem 13.9. □

Within this work we are mostly concerned with the case of $m = k = 1$ and $m = k = 2$, that is either we have a one-dimensional stochastic differential equation driven by one Brownian motion or a system of two equations driven by two Brownian motions.

In either case, an analytic solution of the equations might not be possible such that we have to approximate the solution numerically. For this purpose a variety of discretization schemes are available (see for example Kloeden and Platen [59] or Kloeden, Platen and Schurz [60]). In our context the *Euler–Maruyama scheme* is of sufficient accuracy. Given a stochastic differential equation (1.3), the scheme approximates the process X on a time grid $0 = t_0 < t_1 < \ldots < t_N = T$ by setting $X_{t_0} = \zeta$ and then proceeding for $\nu = 1, \ldots, N$ via the recursion

$$\begin{aligned} X^i_{t_\nu} &= X^i_{t_{\nu-1}} + \beta_i(t_{\nu-1}, X_{t_{\nu-1}})(t_\nu - t_{\nu-1}) + \\ &\quad + \sum_{j=1}^{k} \gamma_{ij}(t_{\nu-1}, X_{t_{\nu-1}})(W^j_{t_\nu} - W^j_{t_{\nu-1}})\;\; \forall\, i = 1, \ldots, m. \end{aligned} \tag{1.4}$$

Coupled with a Monte Carlo discretization of appropriate expectation functionals, the Euler–Maruyama scheme will be used for the approximation of stochastic optimization problems as well as the simulation of hedge error distributions in financial market applications. In the next section we establish the basis for these applications by introducing some fundamental concepts of mathematical finance.

1.2 Fundamentals of Mathematical Finance

Based on the notion of stochastic differential equations we are now able to define the dynamics of financial market models and to introduce several basic concepts of mathematical finance. Note that a more detailed discussion of the presented topics can be found in Elliott and Kopp [31] or Karatzas and Shreve [58]. In the following we consider a general financial market model as defined below.

Definition 1.6. Let $(\Omega, \mathcal{F}, P)$ be a given probability space, $I = [0, T]$ a time index set and $W = (W_t^1, \ldots, W_t^k)_{t \in I}$ a multidimensional Brownian motion with correlation matrix Λ equal to the identity matrix. Further let $\left(\mathbb{F}^W\right)^*$ be the augmented filtration generated by W. Then the $(1 + m, k)$ market model $\mathcal{M}$ is defined by the processes $B = (B_t)_{t \in I}$ and $S = (S_t^1, \ldots, S_t^m)_{t \in I}$ solving the set of stochastic differential equations

$$\begin{aligned} dB_t &= r_t B_t dt, \; B_0 \in (0, \infty), \\ dS_t^i &= \mu_t^i S_t^i dt + S_t^i \sum_{j=1}^{k} \sigma_t^{ij} dW_t^j, \; S_0^i \in (0, \infty), \; i = 1, \ldots, m. \end{aligned} \tag{1.5}$$

Here $(r_t)_{t \in I}$, $(\mu_t^i)_{t \in I}$ and $(\sigma_t^{ij})_{t \in I}$ are measurable processes satisfying the conditions (1.2) such that the integrals in the differential equations are well defined.

In the definition above B describes the dynamic evolvement of a bond price with interest rate r_t, whereas S represents risky assets driven by the stochastic nature of the Brownian motions and the drift μ_t^i. The term $\sigma_{i,t} = (\sum_{j=1}^{k} (\sigma_t^{ij})^2)^{1/2}$ is called the *total volatility* of asset S_t^i. Note that the assumptions in the definition are sufficient to guarantee an almost surely unique solution of the stochastic differential equations.

For our purpose, we are mostly interested in the Black–Scholes as well as stochastic volatility models which are special cases of the general $(1 + 1, 1)$ and $(1 + 1, 2)$ models, respectively. In particular the $(1 + 1, 2)$ model contains the well-known Heston stochastic volatility model and the Stein-Stein model.

In these models our goal is to hedge an exotic option by choosing an appropriate *trading strategy*. We denote such a strategy by $\phi = (\phi_t^0, \phi_t^1, \ldots, \phi_t^m)$, where ϕ_t^0 denotes the units of the bond and ϕ_t^i the units of asset S^i in the portfolio at time t. The value of such a portfolio strategy at time t is hence given by

$$\Pi_t(\phi) = \phi_t^0 B_t + \sum_{i=1}^{m} \phi_t^i S_t^i.$$

A trading strategy ϕ is called *self-financing* if and only if any changes in the portfolio value $\Pi_t(\phi)$ result entirely from net gains (or losses) realized on the investments. Mathematically this requirement is expressed by

$$d\Pi_t(\phi) = \phi_t^0 dB_t + \sum_{i=1}^{m} \phi_t^i dS_t^i,$$

where we neglected the technical assumptions assuring that the integrals in this equation are well defined. The set of all self-financing trading strategies satisfying the corresponding integrability conditions shall be denoted by SF. Among these strategies ϕ so-called arbitrage strategies are of particular interest.

Definition 1.7. A trading strategy $\phi = (\phi_t^0, \phi_t^1, \ldots, \phi_t^m)_{t \in I}$ is an arbitrage strategy if and only if it is self-financing and satisfies the conditions

$$\Pi_0(\phi) = 0, \quad \Pi_T(\phi) \geq 0 \; (a.s.), \quad P(\Pi_T(\phi) > 0) > 0. \tag{1.6}$$

The financial market model $\mathcal{M}$ is said to be arbitrage-free in a given set of trading strategies $\mathcal{S}$ if there exists no $\phi \in \mathcal{S}$ satisfying (1.6).

Hence an arbitrage strategy is a strategy which allows riskless profits in the financial market model $\mathcal{M}$. Note that in addition to strategies satisfying (1.6) also those ϕ fulfilling $\Pi_0(\phi) < 0$, $\Pi_T(\phi) \geq 0$ (a.s.) imply arbitrage opportunities, because we can set $\psi_t^0 = \phi_t^0 - \Pi_0(\phi)/B_0$, $\psi_t^i = \phi_t^i$, $i = 1, \ldots, m$, which in turn implies $\Pi_0(\psi) = 0$ and $\Pi_T(\psi) > 0$ (a.s.). Thus the self-financing trading strategy ψ is an arbitrage strategy.

The notion of no arbitrage is closely related to the concept of *equivalent martingale measures*. This set of measures is defined by

$$\begin{aligned} \mathbb{P} = \{Q \, : \, Q \text{ probability measure on } \mathcal{F}_T, (Q(N) = 0 \Longleftrightarrow P(N) = 0) \; \forall \, N \in \mathcal{F}_T, \\ (S_t^i/B_t)_{t \in I} \text{ is an } \left(\left(\mathbb{F}^W\right)^*, Q\right) - \text{martingale}, \; i = 1, \ldots, m\}. \end{aligned}$$

In analogy we define a localized version $\mathbb{P}_{\text{loc}}$ of this set by only requiring that the process $(S_t^i/B_t)_{t \in I}$ is a local martingale instead of a martingale under Q. Given this set, it is possible to prove the well-known no arbitrage theorem.

Theorem 1.8. *If* $\mathbb{P}_{\text{loc}} \neq \emptyset$ *then the financial market model* $\mathcal{M}$ *is arbitrage-free in the sets of trading strategies* SF_1 *and* $\cup_{Q \in \mathbb{P}_{\text{loc}}} \mathrm{SF}_2(Q)$, *where*

$$\begin{aligned} \mathrm{SF}_1 &= \{\phi \in \mathrm{SF} \, : \, \exists c = c(\phi) \in \mathbb{R} \textit{ such that } \Pi_t(\phi)/B_t \geq c \; (a.s.) \; \forall \, t\} \\ \mathrm{SF}_2(Q) &= \left\{\phi \in \mathrm{SF} \, : \, (\Pi_t(\phi)/B_t)_{t \in I} \textit{ is an } \left(\left(\mathbb{F}^W\right)^*, Q\right) - \textit{martingale}\right\}. \end{aligned}$$

Proof. See Karatzas and Shreve [58], Chapter 1, Theorem 4.2. □

If a market model would allow riskless profits in the form of arbitrage strategies, finding a correct price of a given financial instrument would be virtually impossible. Hence we assume in the following the absence of arbitrage-strategies in the set $\cup_{Q \in \mathbb{P}_{\text{loc}}} \mathrm{SF}_2(Q)$ of our financial market model $\mathcal{M}$ by requesting $\mathbb{P}_{\text{loc}} \neq \emptyset$.

Based on these self-financing trading strategies our goal is now to price and hedge an $\mathcal{F}_T$-measurable *European claim* $C \geq 0$, $C/B_T \in L^1(Q)$, by searching a trading strategy ϕ which replicates this claim as good as possible in all market scenarios. In the ideal case there exists an *exact hedging strategy* for C, that means

$$\exists \phi \in \cup_{Q \in \mathbb{P}_{\text{loc}}} \mathrm{SF}_2(Q) \text{ such that } \Pi_T(\phi) = C \; (a.s.).$$

But a self-financing trading strategy ϕ which exactly replicates the claim at time T must have the same value as C at time $t = 0$, because otherwise arbitrage strategies would exist. Hence it is an intuitive idea to define the fair price of C at time $t = 0$ as

$$\pi(C) := \inf\left\{\Pi_0(\phi) \,:\, \Pi_T(\phi) = C \ (a.s.),\ \phi \in \cup_{\mu\in\mathbb{P}_{\mathrm{loc}}} \mathrm{SF}_2(\mu)\right\}.$$

The question is how to compute this price. Again, the equivalent martingale measures play an important role in this context.

Theorem 1.9. *Suppose that there exists an equivalent martingale measure $Q \in \mathbb{P}_{\mathrm{loc}}$ and a trading strategy $\phi \in \mathrm{SF}_2(Q)$ such that $\Pi_T(\phi) = C$ (a.s.). Then the unique price of C is given by*

$$\pi(C) = \sup_{\mu\in\mathbb{P}_{\mathrm{loc}}} E_\mu\left(\frac{B_0}{B_T}C\right) = E_Q\left(\frac{B_0}{B_T}C\right) = E_Q\left(e^{-\int_0^T r_t dt}C\right). \tag{1.7}$$

Furthermore, for exact hedging strategies $\phi \in \mathrm{SF}_2(Q)$, $Q \in \mathbb{P}_{\mathrm{loc}}$, the value process $V_t := \Pi_t(\phi)$ of claim C is independent of ϕ, Q and given by

$$V_t = B_t E_Q\left(\frac{C}{B_T}\middle|\, \mathcal{F}_t\right) = E_Q\left(e^{-\int_t^T r_s ds}C \,\middle|\, \mathcal{F}_t\right) \ (a.s.), \quad t \in I. \tag{1.8}$$

Proof. See Karatzas and Shreve [58], Chapter 5, Theorem 8.1. □

From a financial point of view the statement of Theorem 1.9 simply means, that in an arbitrage-free financial market model the prices and value processes of all exact hedging strategies coincide and hence imply a unique price for the target claim C under consideration.

However, in general financial market models an exact hedging strategy for C may not exist. In this situation of an *incomplete market* it is not possible to derive a unique price for C by arbitrage arguments as before. But as it turns out, we can still derive an upper bound $\overline{\pi}(C)$ for the price of the claim C by minimizing the portfolio cost $\Pi_0(\phi)$ over the set of all self-financing trading strategies ϕ *super-replicating* C in the sense of $\Pi_T(\phi) \geq C$ (a.s.). In analogy one can derive a lower bound $\underline{\pi}(C)$ by maximizing $\Pi_0(\phi)$ over the space of all *sub-replicating* strategies, that is $\Pi_T(\phi) \leq C$ (a.s.). The result of this process is a price interval $[\underline{\pi}, \overline{\pi}]$ for the value of C at time $t = 0$ instead of a unique price as before. Note that the fair price of C has to lie within these bounds, because otherwise arbitrage strategies would exist. The difference $\overline{\pi} - \underline{\pi}$ is called the *bid-ask-spread* and reflects the idea that $\overline{\pi}$ is the highest price that a buyer is willing to pay for C, whereas $\underline{\pi}$ is the lowest price for which a seller is willing to sell C.

As in the exact hedge case, the prices $\underline{\pi}, \overline{\pi}$ can also be related to equivalent martingale measures (see Karatzas and Shreve [58], Chapter 5, Theorem 6.2) under suitable assumptions such that the interval of all possible prices is given by

$$[\underline{\pi}, \overline{\pi}] = \left[\inf_{\mu\in\mathbb{P}_{\mathrm{loc}}} E_\mu\left(\frac{B_0}{B_T}C\right),\ \sup_{\mu\in\mathbb{P}_{\mathrm{loc}}} E_\mu\left(\frac{B_0}{B_T}C\right)\right].$$

In particular every equivalent martingale measure $Q \in \mathbb{P}_{loc}$ defines a potentially different price $E_Q(\frac{B_0}{B_T}C) \in [\underline{\pi}, \overline{\pi}]$ and a value process $V_t = V_t(Q)$ in the form of (1.8). The question is then which measure Q should be chosen for the pricing of the claim C. In practice the correct Q is determined by the market in the sense of a calibration of the financial market model based on market data.

Further note that the price process (1.8) is arbitrage-free for every $Q \in \mathbb{P}_{loc}$, that is the market model $\mathcal{M}$, extended by the additional tradable asset V, is arbitrage-free in the set of trading strategies SF_1 of the extended model. This fact will be used frequently throughout the book to extend a $(1+1, 1)$ or $(1+1, 2)$ model with several tradable standard calls.

In addition we can reformulate the dynamics (1.5) of the financial market model stated in Definition 1.6 under an equivalent martingale measure $Q \in \mathbb{P}_{loc}$ as

$$\begin{aligned} dB_t &= r_t B_t dt, \; B_0 \in (0, \infty) \\ dS_t^i &= r_t S_t^i dt + S_t^i \sum_{j=1}^{k} \sigma_t^{ij} d\widetilde{W}_t^j, \; S_0^i \in (0, \infty), \; i = 1, \dots, m, \end{aligned}$$

where $\widetilde{W}$ is a Brownian motion obtained by applying the Girsanov theorem (see for example Elliott and Kopp [31], Theorem 7.2.3). This representation of the financial market model $\mathcal{M}$ can be interpreted as a risk-neutral market in which the risky assets S_t^i evolve with the same drift as the bond position. Furthermore, these equations can be discretized numerically with the Euler–Maruyama scheme (1.4) to approximate the price $E_Q(\frac{B_0}{B_T}C)$ by a Monte Carlo-based estimator.

But this is only one possibility to numerically compute the price. Under suitable assumptions it is also possible to derive a pricing partial differential equation which can be solved with methods that are more efficient than Monte Carlo-based algorithms (if the dimension of the problem is not too large). In some cases, it is even possible to analytically solve the partial differential equation. For example, in the Black–Scholes case the price $C(t, s)$ of a standard Call $C = (S_T - K)^+$ with strike K and maturity T satisfies the parabolic differential equation

$$\begin{aligned} C_t + \tfrac{1}{2}\sigma^2 s^2 C_{ss} + (r - \delta) s C_s - rC = 0 & \quad (t, s) \in (0, T) \times (0, \infty) \\ C(T, s) = (s - K)^+ & \quad s \in (0, \infty) \\ C(t, 0) = 0 & \quad t \in (0, T_i), \end{aligned} \tag{1.9}$$

where σ denotes the volatility, r the risk-free rate and δ the dividend yield. This differential equation is solved exactly by the famous *Black–Scholes formula* (see for example Zhang [89]), given the cumulative normal distribution $\Phi(\cdot)$:

$$C(t,s) = se^{-\delta(T-t)}\Phi\left(\frac{\log(\frac{s}{K}) + (r-\delta+\frac{\sigma^2}{2})(T-t)}{\sigma\sqrt{T-t}}\right)$$
$$-Ke^{-r(T-t)}\Phi\left(\frac{\log(\frac{s}{K}) + (r-\delta-\frac{\sigma^2}{2})(T-t)}{\sigma\sqrt{T-t}}\right) =: V(t,s,\sigma,r). \tag{1.10}$$

It is easy to prove that the mapping $\sigma \mapsto V(t,s,\sigma,r)$ is injective and monotonically increasing. Thus, given a price $\pi_{\text{market}}(C)$ of the call observed in the market, it is possible to compute a unique σ^* (if all other parameters are fixed) solving the nonlinear equation $V(t,s,\sigma^*,r) = \pi_{\text{market}}(C)$. This σ^* is called the *implied volatility* of the call and is the unit traders typically use to quote call prices. The advantage of this unit is that the price can be quoted independently of the stock price S_0 at time $t = 0$.

In case the target option under consideration is not a call, traders prefer to quote prices relative to the current stock price S_0. This idea will be used throughout this work. For example we will quote the price € 31.35 of a barrier option, given a stock price S_0 of € 2750 as $31.35/2750 = 0.0114 = 1.14\%$. In addition we may frequently use the financial unit of one *basis point* which is defined as 0.01% of the stock price at time $t = 0$.

Finally, we briefly mention that in practice options are usually hedged dynamically according to their sensitivities with respect to market and model parameters. Mathematically these *sensitivities* are given by the derivatives of the value of the option with respect to the market and model parameters. Without loss of generality we define these sensitivities for the special case of a call with price function $V(t,s,\sigma,r)$ in the Black–Scholes model as defined in (1.10). In this case the most important sensitivities are given by

$$\text{Delta } \frac{\partial C}{\partial s}, \quad \text{Gamma } \frac{\partial^2 C}{\partial s^2}, \quad \text{Vega } \frac{\partial C}{\partial \sigma}, \quad \text{Theta } \frac{\partial C}{\partial t}, \quad \text{Rho } \frac{\partial C}{\partial r}.$$

The dynamic hedge portfolio is then adjusted during the lifetime of the option such that selected Greeks of the target option are matched by the hedge portfolio.

1.3 Optimization Basics

The focus of this work is the identification of optimal super-replication strategies by means of optimization. Accordingly, this section introduces the optimization concepts needed in later chapters. Note that the presented topics are covered in more detail in Goberna and Lopez [46], Hettich and Kortanek [51], Hettich and Zenke [52] as well as Rockafellar [77].

In the following we are interested in the analysis of the optimization problem

$$\min_x f(x) \quad \text{s.t.} \quad x \in X, \tag{1.11}$$

where $f : \mathbb{R}^m \to \mathbb{R}$ is a continuous, convex function and $\emptyset \neq X \subset \mathbb{R}^m$ is a convex set. An optimization problem satisfying these conditions is said to be *convex*. The set X is called the *feasible set* and contains *feasible points* $x \in X$.

The vector $0 \neq d \in \mathbb{R}^m$ is a *direction of recession* of the convex set X if and only if $x + \lambda d \in X$ for every $\lambda \geq 0$ and $x \in X$. In analogy, a vector $0 \neq d \in \mathbb{R}^m$ is a direction of recession of f if $f(x + \lambda d) \leq f(x)$ for all $\lambda \geq 0$ and $x \in X$. Equipped with these definitions, it is possible to prove the following existence theorem for optimization problem (1.11).

Theorem 1.10. *Let $f : \mathbb{R}^m \to \mathbb{R}$ be a continuous, convex function, and let $X \subset \mathbb{R}^m$ be a nonempty closed convex set over which f is to be minimized. If f and X have no direction of recession in common, then f attains its infimum on X, that means there exists an $x^* \in \mathbb{R}^m$ such that $f(x^*) = \inf_{x \in X} f(x)$. In the case where X is polyhedral, f achieves its infimum on X under the weaker assumption that every common direction of recession of f and X is a direction in which f is constant.*

Proof. Note that the function f is proper, that is its domain $\mathbb{R}^m$ is nonempty and the function is finite on this domain. Combined with the continuity of f Theorem 27.3 in Rockafellar [77] yields the desired result. □

As it turns out, the optimization problem of statically hedging a barrier option is a special case of problem (1.11). To be more precise, the static hedging approach leads to a linear semi-infinite optimization problem of the form

$$\min_x c^{\mathrm{T}} x \quad \text{s.t.} \quad a(z)^{\mathrm{T}} x \geq b(z) \quad \forall z \in Z, \tag{1.12}$$

where $Z \subset \mathbb{R}^k$, $|Z| = \infty$, is a given set, $a : Z \to \mathbb{R}^m$, $b : Z \to \mathbb{R}_+$ are suitable mappings and $0 \neq c \in \mathbb{R}^m$. Clearly, the function $f(x) = c^{\mathrm{T}} x$ is a continuous linear function and the feasible set X, implicitly defined by the infinite number of constraints $a(z)^{\mathrm{T}} x \geq b(z)$, is convex, closed, but unbounded (note that $b : Z \to \mathbb{R}_+$, so $b \geq 0$). Hence we can apply Theorem 1.10 to this special case and obtain the following result.

Corollary 1.11. *Suppose that the feasible set of optimization problem (1.12) is non-empty. If there is no vector d satisfying the conditions*

$$d \neq 0, \quad c^{\mathrm{T}} d \leq 0, \quad a(z)^{\mathrm{T}} d \geq 0 \; \forall z \in Z, \tag{1.13}$$

then a solution of optimization problem (1.12) exists.

Proof. It is sufficient to show that the conditions (1.13) precisely characterize a common direction of recession of the objective and the constraints of problem (1.12). But this can be observed easily, because

$$c^{\mathrm{T}}(x + \lambda d) = c^{\mathrm{T}}x + \lambda c^{\mathrm{T}}d \le c^{\mathrm{T}}x \; \forall\, x \in X,\ \lambda \ge 0 \iff c^{\mathrm{T}}d \le 0$$
$$a(z)^{\mathrm{T}}x + \lambda a(z)^{\mathrm{T}}d \ge b(z) \; \forall\, x \in X,\ z \in Z,\ \lambda \ge 0 \iff a(z)^{\mathrm{T}}d \ge 0 \; \forall\, z \in Z.$$

Hence the claim follows by applying Theorem 1.10. □

Due to Corollary 1.11 the existence of a solution of problem (1.12) can be proven by showing that the only vector satisfying the inequalities in (1.13) is the vector $d = 0$. Furthermore, it is possible to analyze the set of optimal solutions in more detail.

Theorem 1.12. *Assume that the feasible set of problem (1.12) is nonempty. Then the following two statements are equivalent:*

i) There is no vector d satisfying the conditions (1.13).
ii) For all $\bar{x} \in X$ the level sets $\{x \in X : c^{\mathrm{T}}x \le c^{\mathrm{T}}\bar{x}\}$ are compact and the set of optimal solutions of problem (1.12) is nonempty, convex and compact.

Proof. See Hettich and Zencke [52], Theorem 3.2.7. □

To solve optimization problem (1.12) numerically, it is an intuitive idea to replace the infinite number of constraints in that problem by a finite grid $M \subset Z$, $|M| < \infty$ which leads to the discretized problem

$$\min_x c^{\mathrm{T}}x \quad \text{s.t.} \quad a(z)^{\mathrm{T}}x \ge b(z) \quad \forall\, z \in M. \tag{1.14}$$

Clearly, a solution x_d^* of this problem is in general infeasible for the original problem (1.12) and the optimal function values satisfy $c^{\mathrm{T}}x_d^* \le c^{\mathrm{T}}x^*$ for a given solution x^* of (1.12). However, we can successively refine the mesh M and hope that the solutions x_d^* of the discretized problems converge to the set of optimal solutions of the original problem. The next theorem shows that the desired convergence can be achieved.

Theorem 1.13. *Suppose that the feasible set of optimization problem (1.12) is nonempty, the set Z is compact and that one of the equivalent conditions in Theorem 1.12 holds. Then for each $\epsilon > 0$ there exists a $\delta > 0$ such that for every discrete subset $M \subset Z$, $|M| < \infty$, satisfying*

$$\Delta(M) = \max_{z \in Z} \min_{z_1 \in M} \|z - z_1\|_2 \le \delta,$$

the discretized problem (1.14) has a solution x_d^* and for each of these solutions there exists an x^* solving (1.12) with $\|x_d^* - x^*\|_2 \le \epsilon$.

Proof. See Hettich and Zencke [52], Theorem 3.2.10. □

Thus, as the mesh M gets more and more dense in Z, i.e. $\Delta(M) \to 0$, the solutions of the discretized problems (1.14) converge to a solution of the semi-infinite optimization problem (1.12). The resulting subproblems (1.14) are (potentially large-scale) linear programming problems which can be solved by available standard solvers. For a survey of available linear programming solvers see for example Wright [88].

However, due to the curse of dimensionality (see Bellman [8]) such a mesh refinement is only possible if the dimension of the parameter space Z is small. If the dimension is too large, alternative solution methods have to be applied which prevent an exponential growth of the number of nodes in the mesh. For example this can be achieved by applying the exchange method of semi-infinite programming to our particular problem. Instead of uniformly or locally refining the mesh in all parameter dimensions, this method computes the worst case constraint violation of a given solution x_d^* of (1.14) by solving the optimization problem

$$\min_z a(z)^{\mathrm{T}} x_d^* - b(z) \quad \text{s.t.} \quad z \in Z. \tag{1.15}$$

In general this optimization problem is nonlinear and hence has to be solved by appropriate deterministic optimization methods (see e.g. Bonnans et al. [12]). If the optimal function value is greater than $-\text{TOL}$, where $\text{TOL} > 0$ is a prespecified convergence tolerance, the desired degree of accuracy is reached. Otherwise the optimal solution z^* of (1.15) is added to the grid M and problem (1.14) is solved again. This procedure is repeated recursively until the termination criterion is fulfilled. In Section 5.2 we provide a detailed algorithm implementing this idea and prove the convergence of the method for our particular hedging problem. Further we illustrate for an example with a six-dimensional parameter space Z, that the number of nodes in M stays relatively small.

In the remainder of this section we majorly focus on an example of problem (1.12) which will be particularly important in the context of static super-replication strategies.

Definition 1.14. Assume that μ is a strictly positive measure on the compact set $Z \subset \mathbb{R}^k$ and the associated Borel sigma algebra $\mathcal{B}(Z)$. Further let $a_1, \dots, a_m : Z \to \mathbb{R}$ be continuous functions which are linearly independent. Then, for the continuous function $b : Z \to \mathbb{R}$, the one-sided L^1-approximation problem is defined as the optimization problem

$$\min_{x \in \mathbb{R}^m} \int_Z \left(\sum_{i=1}^m x_i a_i(z) - b(z) \right) d\mu(z) \quad \text{s.t.} \quad \sum_{i=1}^m x_i a_i(z) \geq b(z) \; \forall \, z \in Z. \tag{1.16}$$

Note that in the definition above a strictly positive measure on Z with the associated Borel sigma algebra is defined as a measure satisfying

$$\int_Z h(z) d\mu(z) > 0 \; \forall \text{ continuous functions } h : Z \to \mathbb{R}, \; h \geq 0, \; h \neq 0.$$

Obviously, problem (1.16) is of the form (1.12) if we choose $c_i = \int_Z a_i(z)d\mu(z)$. But then (1.13) implies that a common direction of recession d of the objective and the constraints has to satisfy

$$d \neq 0, \quad \int_Z \sum_{i=1}^{m} d_i a_i(z) d\mu(z) \leq 0, \quad \sum_{i=1}^{m} d_i a_i(z) \geq 0 \; \forall \, z \in Z.$$

Due to the positivity of the measure we can thus conclude that $\sum_{i=1}^{m} d_i a_i(z) = 0$ which can only occur if $d = 0$, because the functions a_i are linearly independent. Hence Theorem 1.12 shows that under the given assumptions a solution of the L^1-approximation problem exists and that the set of solutions is convex and compact.

If problem (1.16) additionally satisfies the *slater condition*, that is there exists a vector $x \in \mathbb{R}^m$ such that $a(z)^{\mathrm{T}} x > b(z) \; \forall \, z \in Z$, then it is possible to show that the strong duality theorem holds.

Theorem 1.15. *If the feasible set of the one-sided L^1-approximation problem (1.16) is nonempty and the slater condition is satisfied, then the dual problem*

$$\begin{aligned} &\max_{\nu} \int_Z b(z) d\nu(z) \\ &\text{s.t. } \nu : \mathcal{B}(Z) \to [0, \infty) \text{ measure on the Borel sigma algebra } \mathcal{B}(Z) \\ &\int_Z a_i(z) d\nu(z) = \int_Z a_i(z) d\mu(z) \end{aligned} \tag{1.17}$$

is solvable and the optimal function values of (1.16) and (1.17) coincide. Furthermore, there exists an optimal solution ν^* of the dual problem (1.17) which is the weighted sum of at most m Dirac measures $\delta_{\{z_l\}}$, i.e. $\nu^* = \sum_{l=1}^{r} w_l \delta_{\{z_l\}}$, $r \leq m$, $w_l \geq 0$, where $z_l \in Z$ are points in which the constraints of (1.16) are active for a given optimal solution x^* of (1.16).

Proof. See Hettich and Zencke [52], Examples 2.3.8, 3.2.16 and Theorem 3.2.13. □

Hence in the particular example of a one-sided L^1-approximation problem with continuous functions a_i and b the theory of semi-infinite optimization delivers very interesting results regarding the dual problem. Unfortunately, in our application the function b turns out to be discontinuous such that the dual problem cannot be derived by applying the theorem above directly. However, in Section 4.2 we present a way to circumvent this problem and still obtain a nice duality result.

We conclude this section with a brief remark on two other optimization problems which occur in this book. Within the context of the robustification of a static trading strategy semidefinite as well as second order cone problems arise. A semidefinite programming problem in standard form is given by (see for example Vandenberghe and Boyd [86])

$$\min_{x \in \mathbb{R}^m} c^{\mathrm{T}} x \quad \text{s.t.} \quad \sum_{i=1}^{m} x_i A_i - B \succeq 0,$$

where $c \in \mathbb{R}^m$ is a given vector, $A_i, B \in \mathbb{R}^{k \times k}$ are symmetric matrices and $Y \succeq 0$ is equivalent to Y being positive semidefinite. The semidefinite program above also contains the interesting class of second order cone problems

$$\min_{x^i \in \mathbb{R}^{m_i}} \sum_{i=1}^{k} (c^i)^{\mathrm{T}} x^i \quad \text{s.t.} \quad \sum_{i=1}^{k} A_i x^i = b, \quad x_1^i \geq \|(x_2^i, \ldots, x_{m_i}^i)^{\mathrm{T}}\|_2 \ \forall\, i$$

as a special case (see Alizadeh and Goldfarb [1]). Here c^i, b and A_i are real vectors and matrices of appropriate dimension and the vector $x^i \in \mathbb{R}^{m_i}$ is written componentwise as $x^i = (x_1^i, x_2^i, \ldots, x_{m_i}^i)^{\mathrm{T}}$, $i = 1, \ldots, k$. Both classes of problems are special convex programs of the form (1.11) and can be solved efficiently by interior point methods (see Ben-Tal and Nemirovski [9]).

The presented optimization concepts as well as the fundamentals of probability theory and financial mathematics described in the previous sections form the basis of the analysis and derivations to come. A summary of the symbols and abbreviations introduced in this chapter and throughout the book can be found on page 192.

2 Static Hedging of Barrier Options

2.1 Motivation

The growing demand for derivative products in the financial market industry has led to the development of a large variety of structured products. This growth is also reflected in the recent developments at the EUREX, the world's leading futures and options market for Euro denominated derivative instruments. In 2008, the trading volume in this market exceeded 2.1 billion contracts. At the same time equity index derivatives became the largest trading segment and reached more than 1 billion contracts. Among those, derivatives on the Dow Jones EURO STOXX 50 index were the largest single product with 432 million futures and 401 million options.

Along with this rapid growth of the option markets goes hand in hand an increasing risk of the banks issuing these products to the customers. This risk stems from the fact that the bank, by selling the product to the customer at time $t = 0$, enters the obligation of an uncertain future payment at maturity $T > 0$. This payment is based on the future performance of a so-called underlying, usually the stock, from time $t = 0$ to time $t = T$.

To reduce the risk of the uncertain future payment traders immediately set up a portfolio of financial instruments which hopefully matches the payoff of the sold product closely in all possible states of the market. Usually this *hedge portfolio* is dynamically adjusted during the lifetime of the option according to the sensitivities (*Greeks*) of the sold product with respect to the stock price (*delta*) or chosen model parameters (*vega* etc.). This standard hedging method works sufficiently well if the underlying is liquidly traded and the Greeks are not too sensitive with respect to movements of the market. Only if the latter two assumptions are satisfied, traders can quickly react to price changes of the underlying and adjust their portfolio positions according to the changed Greeks.

However, as the recent financial market crisis has shown, market liquidity can dry out quickly, which adversely affects the necessary dynamic adjustment of the hedge. Moreover, in the case of barrier options, the Greeks behave very differently from those of standard options. For example the delta of an up-and-out call varies wildly if the underlying is close to the barrier and the option close to maturity. This behavior is caused by a discontinuity of the barrier option payoff illustrated in Figure 2.1.

The gap between the initial and boundary condition of the Black–Scholes partial differential equation leads to an extreme slope of the solution close to the corner (D, T). Hence the delta as well as its derivative with respect to the stock price (*gamma*) take large negative values close to (D, T) even approaching minus infinity. Consequently, dynamic hedging of barrier options based on sensitivities is virtually impossible in the worst case of a barrier hit close to maturity.

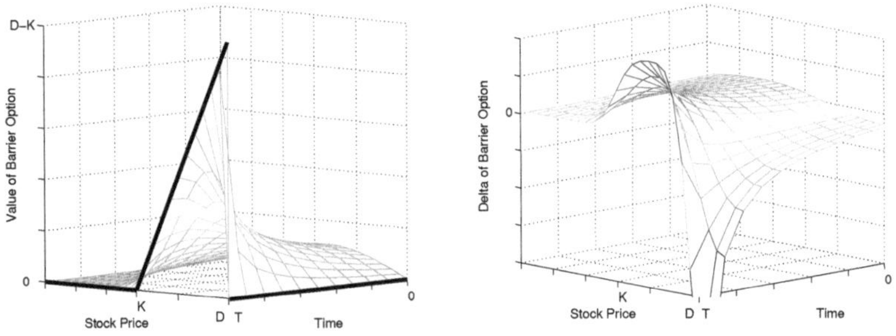

Figure 2.1: Value (left) and delta (right) of an up-and-out call with maturity T, strike K and barrier $D > K$

2.2 Semi-Static Hedging Strategies

The problems of dynamic hedging have led to the development of so-called static trading strategies (see Section 2.3 for a detailed review of the literature). This approach aims at setting up a portfolio of standard options which matches the payoff of the target barrier option without any dynamic adjustment of the portfolio positions. To be precise the portfolio is kept constant until either a barrier hit occurs or the barrier option expires. In order to characterize these trading strategies more formally, we build upon the following general financial market model.

Assumption 2.1. We consider a financial market model $\mathcal{M}$ with time index set $I = [0, T]$ or $I = \{t_0, t_1, \ldots, t_f\}$, $0 = t_0 < t_1 < \ldots < t_f = T$, probability space $(\Omega, \mathcal{F}, P)$, a bond B, one tradable risky underlying S and European options[1] $C^i, i = 1, \ldots, n$, with maturities $T_i \leq T$ which are all written on the risky underlying S. Let the price processes of the bond, the underlying and the options[2] be given by $(B_t)_{t \in I}$ $(B_t > 0 \; \forall t \in I)$, $(S_t)_{t \in I}$ and $(C_t^i)_{t \in I}, i = 1, \ldots, n$, respectively. $(B_t)_{t \in I}$, $(S_t)_{t \in I}$ and $(C_t^i)_{t \in I}, i = 1, \ldots, n$, shall be adapted to the filtration $(\mathcal{F}_t)_{t \in I}$, $\mathcal{F}_0$ is assumed to be trivial. Furthermore, we impose the regularity assumption $B_t, C_t^i \in L^1(P)$ for all $t \in I$ and $i = 1, \ldots, n$.

In the financial market model $\mathcal{M}$ we consider a knock-out barrier option with maturity T whose terminal payoff can be decomposed into

$$C_{\mathrm{KO}} = H 1_{\{\tau > T\}} + R 1_{\{\tau \leq T\}}, \tag{2.1}$$

where H is an $\mathcal{F}_T$-measurable random variable, R is constant and τ is the stopping time of the first barrier event. For example, for a standard up-and-out call option with

[1] Usually we will consider standard call and put options as well as binary calls and puts.

[2] In case $T_i < T$, the price process of option C^i is extended by $C_t^i = C_{T_i}^i B_t / B_{T_i}$, $t \in (T_i, T]$.

strike K, barrier $D > S_0$ and rebate R, we would have

$$H = (S_T - K)^+, \quad \tau = \inf\{t \in I : S_t \geq D\},$$

where we define $\inf \emptyset := +\infty$. Our goal is to statically hedge barrier options by holding a portfolio of the tradable options C^i, $i = 1, \dots, n$, and the bond B and change the portfolio weights only once at the first barrier event occurring before maturity. More formally, we define these static knock-out trading strategies[3] as follows.

Definition 2.2. Let $\mathcal{M}$ be a market model according to Assumption 2.1. Further let a knock-out barrier option with maturity T and payoff $C_{\mathrm{KO}} = H1_{\{\tau>T\}} + R1_{\{\tau\leq T\}}$ be given, where H is $\mathcal{F}_T$-measurable, R is constant and τ is the stopping time of the first barrier event. Consider a trading strategy $\phi = (\phi_t^0, \phi_t^1, \dots, \phi_t^n)_{t\in I}$. Here ϕ_t^0 denotes the units of the bond B and ϕ_t^i the units of option C^i, $i = 1, \dots, n$, in the trading portfolio at time t. The trading strategy ϕ is called a "static knock-out trading strategy" if there exist constants $\alpha_0, \alpha_1, \dots, \alpha_n \in \mathbb{R}$ such that for all $t \in I$

$$\begin{aligned} \phi_t^0 &= \alpha_0 + \frac{\sum_{i=1}^n \alpha_i C_{\tau\wedge T}^i}{B_{\tau\wedge T}} 1_{\{\tau\leq t\}} \\ \phi_t^i &= \alpha_i 1_{\{\tau>t\}}, \quad i = 1, \dots, n. \end{aligned}$$

The set of all these static knock-out trading strategies shall be denoted by $\mathcal{S}_{\mathrm{KO}}$.

The definition above implies that a static knock-out trading strategy involves buying/selling certain quantities of the bond B and the tradable options $C^1, \dots, C^n$ at the beginning and only changing the hedge portfolio in case the first barrier event occurs before the maturity of the barrier option, which then requires unwinding the hedge portfolio and transferring all the proceeds into the bond position. This behavior is illustrated in Figure 2.2 for the case of an up-and-out call based on two possible stock price paths. We call these strategies static as they involve holding a static portfolio until the first barrier event occurs, after which the portfolio is liquidated. By definition such a strategy is self-financing.

In contrast to dynamic trading strategies, the static hedges specified in Definition 2.2 in particular do not require a continuous rebalancing of the portfolio positions during the lifetime of the barrier option. Consequently, such hedges are less exposed to transaction costs and a dry-out of market liquidity than the traditional dynamic hedges. However, so far it is unclear if the restrictive buy-and-hold nature of static strategies is capable of successfully hedging a barrier option. Moreover, we did not specify yet how to choose the standard options $C^1, \dots, C^n$ as well as the portfolio weights $\alpha_0, \dots, \alpha_n$. The next section will address these questions by presenting the existing approaches in the literature.

[3] In analogy static knock-in trading strategies can be defined to hedge knock-in barrier options.

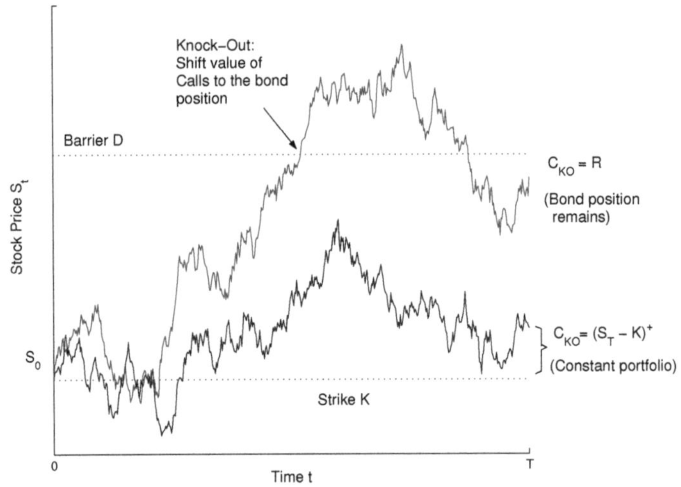

Figure 2.2: Portfolio positions and payoff of an up–and–out call for two possible stock–price paths

2.3 Review of the Static Hedging Literature

Various authors have contributed to the area of static hedging, starting with the pioneering papers of Bowie and Carr [16] and Derman et al. [27]. In the years afterwards the literature on static hedging flourished, leading to several quite different approaches to the topic. In this section we will introduce these approaches, which can in principle be distinguished by the choice of the standard options $C^1, \ldots, C^n$, the model in which the hedge is derived and if the hedge is exact or not.

2.3.1 Strike Spread Hedge

The strike spread static hedging approach aims at replicating the barrier option by a portfolio of standard options with varying strikes, but the same maturity as the target option. Bowie and Carr [16] formally derived the necessary set of strikes for the static strategy described in Definition 2.2 in a Black–Scholes model

$$\begin{aligned} dS_t &= (r-\delta)S_t dt + \sigma S_t dW_t, \; S_0 > 0 \\ dB_t &= rB_t dt, \; B_0 > 0, \end{aligned} \tag{2.2}$$

with zero cost of carry (i.e. $r - \delta = 0$). To illustrate the idea we consider the hedging of an up-and-out call with maturity T, strike K, barrier $D > S_0$ and zero rebates with time-T payoff

$$C_{\text{uo}} = (S_T - K)^+ 1_{\{\tau > T\}}, \quad \tau = \inf\{t \in [0,T] \,:\, S_t \geq D\}. \tag{2.3}$$

Following Bowie and Carr [16] this up-and-out call can be perfectly replicated in model (2.2) satisfying $r - \delta = 0$ with the path-independent payoff

$$f(S_T) = (S_T - K)^+ - \frac{K}{D}\left(S_T - \frac{D^2}{K}\right)^+ - (D-K)\left[2 \cdot 1_{\{S_T \geq D\}} + \frac{1}{D}(S_T - D)^+\right]$$

at time T. A closer look at this payoff reveals that it is simply the combination of a call with the same strike as the barrier option, certain quantities of calls with strike D^2/K or D, and $-2(D-K)$ digital calls with strike D. In terms of Definition 2.2 the static hedging strategy is hence based on positions in three plain vanilla calls and a digital call. Figure 2.3 illustrates the portfolio payoff $f(S_T)$ graphically and underlines the following intuitive explanation of the hedge.

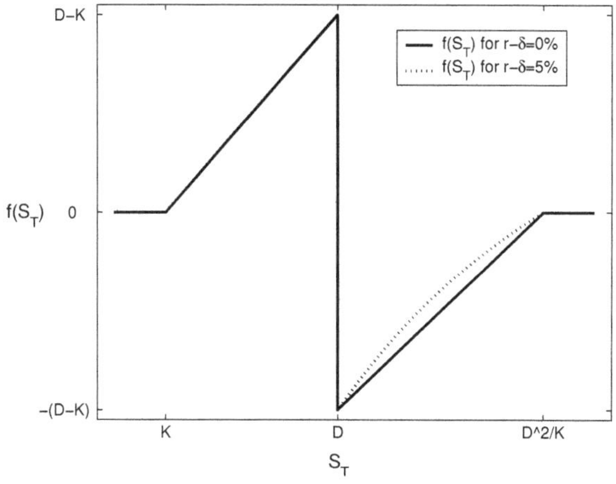

Figure 2.3: Payoff of the exact strike-spread static hedge portfolio for an up-and-out call at time T. For the case $r - \delta = 5\%$ we assumed a Black–Scholes volatility of $\sigma = 20\%$.

On the one hand the call with the same strike as the barrier option perfectly replicates the option if the barrier is not hit ($\tau > T$), since then $C_{\text{uo}} = (S_T - K)^+$ with $S_T < D$ which implies that the calls with strike D^2/K and D as well as the digital calls with strike D expire worthless.

If, on the other hand, the stock price hits the barrier D until time T ($\tau \leq T$) then, according to the static hedging strategy (see Definition 2.2), the standard options are liquidated exactly at the time of the barrier hit with a corresponding stock price $S_\tau = D$. At this time the value of the up-and-out call equals zero, which is the value the hedge needs to replicate. To achieve this, the positive payoff of the standard call with strike K is reflected above the barrier D, which in the sum approximately leads to an integrated payoff of zero.

Bowie and Carr [16] and Carr, Ellis and Gupta [24] formally prove under the assumption of zero drift ($r - \delta = 0$) that the value of the portfolio with final payoff

$f(S_T)$ is zero along the barrier, such that the proposed static hedge indeed yields an exact hedge portfolio for the barrier option.

If the drift of the stock price (2.2) under the risk-neutral measure is nonzero (e.g. $r - \delta = 5\%$), then the probability of the stock price S_T to attain values above the barrier D increases. In this situation the hedge portfolio payoff $f(S_T)$ would no longer have an expected value of zero, but instead lead to hedging losses on the barrier. Consequently the payoff needs to be adjusted to preserve the property of an exact hedge. Carr and Chou [23] showed for this case in the Black–Scholes model (2.2) that the adjusted payoff function

$$f(S_T) = \begin{cases} -\left(\frac{S_T}{D}\right)^z \left(\frac{D^2}{S_T} - K\right)^+ & \text{if } S_T > D \\ (S_T - K)^+ & \text{if } S_T \leq D, \end{cases} \tag{2.4}$$

where $z := 1 - 2(r - \delta)/\sigma^2$, perfectly hedges the up-and-out call. Figure 2.3 graphically illustrates $f(S_T)$ for $r - \delta = 5\%$ and $\sigma = 20\%$. Obviously, the positive drift under the risk-neutral measure is compensated by slightly decreasing the negative area of the reflected payoff function above the barrier D, leading to a portfolio value of zero along D (moving backwards in time). Approximating the resulting nonlinear payoff function with a set of plain vanilla options leads to a nearly exact static hedge based on standard contracts.

These intuitively appealing results can further be extended to the case of a nonconstant, but symmetric volatility function (for details see Carr, Ellis and Gupta [24]). However, even in this general case the assumptions imposed for the derivation of the hedge are violated by most pricing models used in practice. In addition calls with strikes equal to D^2/K (e.g. $[120\%]^2/100\% = 144\%$ of the current stock price) are rarely traded in the market such that it is difficult to set up the hedge in practice.

2.3.2 Calendar Spread Hedge and Extensions

Derman, Ergener and Kani [27] take an alternative approach by exactly replicating the barrier option payoff along the barrier at a finite number of points with standard options of varying maturity (*calendar spread*). To illustrate the idea we consider again the static hedge for an up-and-out call (2.3) with maturity T, strike K, barrier D and zero rebates in the Black–Scholes model (2.2).

Similar to the strike spread approach we first choose the plain vanilla call with the same strike and maturity as the barrier option. This call guarantees a perfect replication of the target option payoff if the barrier D is not hit during the lifetime of the option. However, if the barrier is hit at time $\tau(\omega) < T$ this call has a significant positive value, whereas the up-and-out call expires worthless.

The idea is now to offset this positive value of the call at a discrete set of time points $0 = t_1 < t_2 < \ldots < t_{n_t} < T$ along the barrier by adding calls with varying maturity to the portfolio. First we select a standard call C^{n_t} with maturity T and strike D, and choose its portfolio weight α_{n_t} such that the sum of this call position and the standard call C^{n_t+1} with the same strike and maturity as the barrier option equals zero at time t_{n_t} if the stock price $S_{t_{n_t}}$ equals the barrier D, i.e.

$$\alpha_{n_t} C^{n_t}(t_{n_t}, D) + C^{n_t+1}(t_{n_t}, D) = 0.$$

In this equation (and in the following) $C^i(t_j, D)$ denotes the Black–Scholes value of the ith hedge instrument at time t_j if the underlying S_{t_j} equals the barrier D.

Clearly, if the barrier is not hit during the lifetime of the option, then option C^{n_t} expires worthless and does not adversely affect the perfect replication based on the remaining standard call C^{n_t+1}. In addition the portfolio consisting of both options matches the value of the up-and-out call if the barrier is hit at time t_{n_t}. At time t_{n_t-1}, however, the value of the portfolio is not necessarily zero. Consequently, we add another call C^{n_t-1} with maturity t_{n_t} and strike D to the portfolio, and choose its portfolio weight α_{n_t-1} such that

$$\alpha_{n_t-1} C^{n_t-1}(t_{n_t-1}, D) + \alpha_{n_t} C^{n_t}(t_{n_t-1}, D) + C^{n_t+1}(t_{n_t-1}, D) = 0.$$

Proceeding recursively through time, we can construct a portfolio consisting of $n_t + 1$ calls which satisfies for each $j = 1, \dots, n_t$ the equation

$$\sum_{i=j}^{n_t} \alpha_i C^i(t_j, D) + C^{n_t+1}(t_j, D) = 0. \tag{2.5}$$

Figure 2.4 illustrates graphically how the portfolio value of the resulting static hedge portfolio looks like on the barrier. The time points t_j at which the hedge portfolio exactly matches the payoff of the up-and-out call are visible. At maturity T the value of the hedge portfolio has to bend upwards to $D - K$ since the hedge needs to replicate the final payoff $(S_T - K)^+$. Clearly, the remaining hedge error along the barrier can be reduced further by adding more time points t_j with associated standard calls. In the limit a perfect replication of the up-and-out call can be constructed based on an infinite number of options (see Derman, Ergener and Kani [27] for details).

Compared to the strike spread approach presented in Subsection 2.3.1 the strikes of the options used for the calendar spread hedge need not exceed the barrier D and hence are more likely to be traded in the market. However, a sufficiently accurate calendar spread hedge will consist of a large number of options which makes the strategy again less practical. Nevertheless the idea to match the value of the barrier option along the barrier at a discrete number of times was picked up by various authors who extended the approach beyond the Black–Scholes framework.

Motivated by the analysis of Toft and Xuan [83], who showed that the Derman–Ergener–Kani hedge deteriorates in a stochastic volatility environment, Fink [38] tries to improve the hedge by matching the barrier option payoff on a time-volatility grid

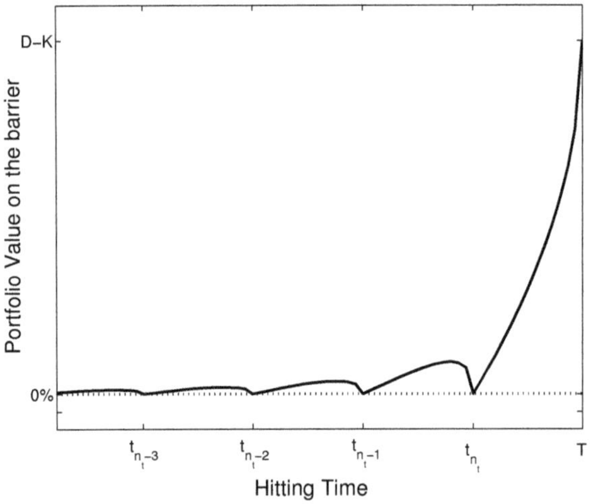

Figure 2.4: Value of the Derman–Ergener–Kani hedge portfolio for an up-and-out call along the barrier for various barrier hit times.

instead of a simple time grid along the barrier. In analogy to the set of equations (2.5) this leads to the conditions

$$\sum_{i=j}^{n_t}\sum_{l=1}^{n_v} \alpha_{il} C^{il}(t_j, D, v_k) + C^{n_t \times n_v + 1}(t_j, D, v_k) = 0, \qquad j = 1, \dots, n_t,\ k = 1, \dots, n_v, \tag{2.6}$$

for a hedge which replicates the up-and-out call in n_v volatility states $v_1, \dots, v_{n_v}$ at each of the n_t time points $0 = t_1 < \dots < t_{n_t}$. In this setting the portfolio weights α_{il} can be computed by recursively solving a linear system of equations for each time point t_j, starting at t_{n_t}.

Unfortunately the required number of hedge instruments to satisfy conditions (2.6) is now $n_t \times n_v + 1$, which easily accumulates to an unmanageable huge number of standard options whose tradability in the market is questionable. Furthermore, as we will see in Subsection 3.3.2, the linear system of equations is ill-conditioned leading to exploding hedge positions and instabilities of the hedge error function.

The size and ill-posedness of the linear system increases even more if we require the hedge portfolio to match the up-and-out call in further market states. This was also observed by Allen and Padovani [2] who studied the behavior of the hedge portfolio if in addition to time and volatility states also n_s different shapes of the volatility surface are considered. In analogy to Fink's [38] findings the large number of $n_t \times n_v \times n_s + 1$ standard options leads to exploding portfolio positions and very instable hedge results which are not suitable for the risk management of exotic options.

To mitigate these drawbacks of the calendar spread approach, Nalholm and Poulsen

[73] propose two important modifications of the method. First the solution of the large linear system of equations is not computed exactly, but instead approximately based on a singular value decomposition. Besides a significant reduction of the size of the hedge positions this approach also reduces the number of hedge instruments to an acceptable level. Secondly, the authors propose to first choose a portfolio of hedge instruments with maturity T which mimics the strike spread approach (see Subsection 2.3.1), and then to offset the residual hedge error by solving the linear systems of type (2.6). In comparison to the original Derman–Ergener–Kani approach the strike spread component of the hedge has a relatively small hedge error on the barrier, such that offsetting this error with additional calendar spread hedges can be achieved with much smaller hedge positions.

However, even though the latter modifications significantly improve the practicability of the calendar spread hedge, the main problem of an exploding size of the linear system and an unknown hedge error between the nodes of the considered market scenario grid is not solved (neglecting potentially significant hedge errors that are introduced by the singular value decomposition).

2.3.3 The Static Mean Square Hedge

One of the main drawbacks of the calendar spread hedge discussed in the previous subsection is that it aims at exactly matching the barrier option payoff in a large number of market scenarios, which in turn requires an equally large number of hedge instruments. To prevent an explosion of the number of hedge instruments the idea arises to give up the notion of an exact hedge and to consider approximate static hedging strategies instead.

An intuitive approach is to find the "best" inexact static strategy by optimization, where the notion of "best" is based on a given measure of the hedge error. For this purpose several measures have been developed in the literature. For example, in a discrete model Dupont [29] chooses to minimize the L^2-hedge error (first introduced by Föllmer and Schweizer [40] for dynamic hedges) in the context of static hedging of barrier options. According to this approach an optimal hedge for a knock-out call (2.1) would be the solution of the optimization problem

$$\min_{\phi \in \mathcal{S}_{\mathrm{KO}}} E_P[\Pi_T(\phi) - C_{\mathrm{KO}}]^2, \tag{2.7}$$

where $\Pi_T(\phi) = \Pi_T((\phi_t^0, \dots, \phi_t^n)_{t \in I}) = \phi_T^0 B_T + \sum_{i=1}^n \phi_T^i C_T^i$ denotes the value of the hedge portfolio at time T for a given strategy $\phi \in \mathcal{S}_{\mathrm{KO}}$ as specified in Definition 2.2.

However, from the point of view of a bank selling the barrier option the symmetric treatment of the hedge error is a conceptual problem, because losses and gains are penalized in the same way. In addition the L^2-hedge error measure has the tendency to produce static hedge portfolios for barrier options with a good average performance, but extreme hedging losses for some low probability events. This conceptual problem will be illustrated in Subsection 3.3.2 in the form of a typical numerical example. Besides this disadvantage the quadratic minimization problem (2.7) is furthermore ill conditioned due to the high correlation of the financial products included in the hedge portfolio. This leads to many quite different portfolios with a similar value of the hedge error measure and usually large hedge positions.

To cope with the latter problem various approaches have been developed in the literature. For example Boyle, Coleman and Li [17] propose the regularization of a cost-objective via a penalty approach subject to bounds on the hedge positions and an upper bound on the mean-square-hedge-error. In [25] Coleman, Li and Patron choose the L^1-error as a risk measure and compare their results to the quadratic hedging approach.

But in all these studies the main problem of a symmetric treatment of the hedge error is not solved. Due to the tendency of static hedges to produce extreme losses with low probability, robust one-sided hedge error measures are needed which explicitly penalize or avoid such losses. From a conservative point of view this leads to the idea of super-replicating the barrier option with static hedging strategies.

2.3.4 Model-Independent Super-Hedges

To prevent potential losses, combining static hedges with the idea of super-replication (see e.g. El Karoui and Quenez [30] for super-replication in the context of dynamic trading strategies) seems to be highly desirable. For the case of an up-and-out call (2.3) a superhedge can be derived from the simple (model-independent) inequality

$$\begin{aligned} C_{\text{uo}} &= (S_T - K)^+ 1_{\{\tau > T\}} \leq (S_T - K)^+ 1_{\{S_T < D\}} \\ &= (S_T - K)^+ - (S_T - D)^+ - (D - K) 1_{\{S_T \geq D\}}. \end{aligned} \tag{2.8}$$

As the latter expression and Figure 2.5 show, the super-replicating payoff can be decomposed into standard calls with strikes K, D and maturity T, as well as $-(D - K)$ European digital calls with payoff $1_{\{S_T \geq D\}}$. Since this payoff (in contrast to the barrier option) is path-independent, we can deduce the model-independent price of the superhedge from current market prices of plain vanilla calls and a call-spread approximation of the digital option. However, based on inequality (2.8) it is immediately clear that this upper bound is very crude since the superhedge can lead to large positive cash flows in case of a barrier hit, whereas the up-and-out call expires worthless. This is also confirmed by comparing the superhedge payoff sketched in Figure 2.5 with the payoff of the static strike-spread hedge (which exactly replicates the up-and-out call under certain assumptions) illustrated in Figure 2.3.

Nevertheless Brown, Hobson and Rogers [19] as well as Neuberger and Hodges [74] are able to prove under the assumption of zero drift and right-continuity of the stock price process that the price of the superhedge (2.8) is the smallest model-independent upper bound for the up-and-out call. In particular the authors prove the existence

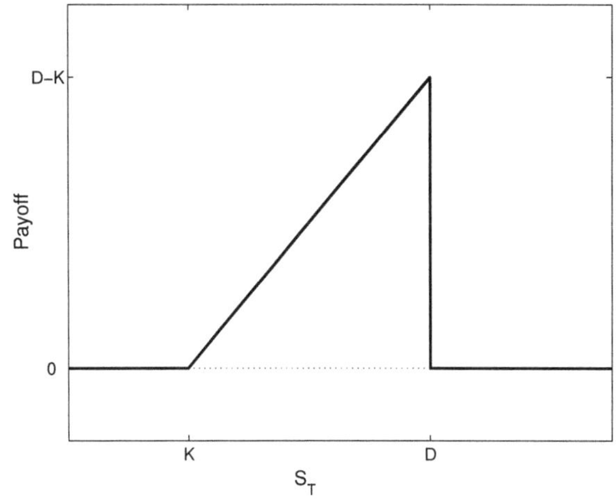

Figure 2.5: Payoff of the model-independent superhedge (2.8) for an up-and-out call.

of a jump stock price model which is consistent with the observed market prices of European options at maturity T, and whose model price of the up-and-out call equals the model-independent price of the superhedge (2.8).

If the additional assumption of continuous stock price paths is imposed, this very conservative bound can be improved based on the inequality

$$\begin{aligned} C_{\text{uo}} &= (S_T - K)^+ 1_{\{\tau > T\}} \\ &\leq \frac{(\beta - K)(D - S_0)}{D - \beta} + \frac{\beta - K}{D - \beta}(S_0 - S_T) + \\ &\quad + \frac{D - K}{D - \beta}\{(S_T - \beta)^+ - (S_T - D)^+\} - (D - K)1_{\{S_T \geq D\}} + \\ &\quad + \frac{\beta - K}{D - \beta}(S_T - D)1_{\{\tau \leq T\}} \end{aligned} \tag{2.9}$$

for an arbitrary constant $\beta \in [K, D)$. The static hedge associated with this inequality consists of setting up a portfolio of the first four summands on the right hand side of (2.9) at time $t = 0$. If the barrier D is hit during the lifetime of the option (i.e. $\tau \leq T$), and consequently the stock price equals the barrier at that time (due to the continuously evolving stock price paths), we enter into a forward transaction with strike exactly equal to the barrier. This transaction corresponds to the last summand in (2.9) and has a price of zero at the time of the barrier hit since we assumed that the stock price process has zero drift.

Figure 2.6 graphically illustrates the right hand side of inequality (2.9) for the two cases of a barrier hit ($\tau \leq T$) and no barrier hit ($\tau > T$) during the lifetime of the option. Obviously the final payoff of the static hedge super-replicates the up-and-out call payoff in both cases.

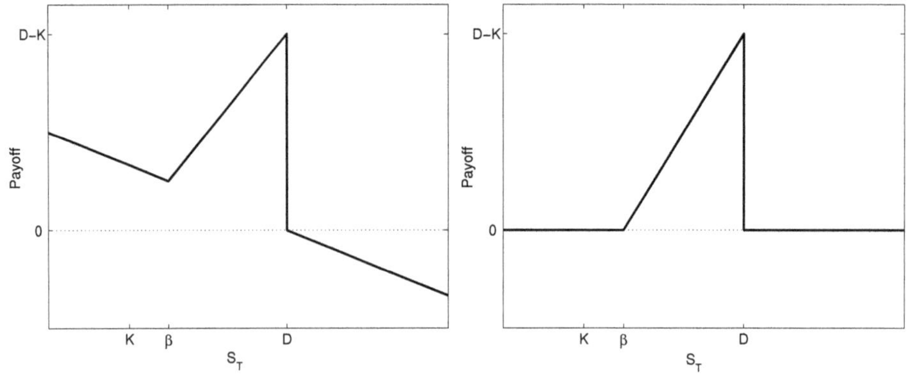

Figure 2.6: Payoff of the model-independent Brown–Hobson–Rogers superhedge at maturity without (left) and with (right) a barrier hit during the lifetime of the option.

Since the constant $\beta \in [K, D)$ was arbitrary, we can choose it optimally in the sense of a minimal price of the super-replication hedge. But even after this optimization the static superhedge (2.9) leads to upper bounds for barrier options which are far too conservative from a practical point of view.

For example, in a Black–Scholes world with zero drift and constant volatility of 20% the Brown–Hobson–Rogers superhedge for an up-and-out call with maturity $T = 1$, strike $K = 100\% = S_0$ and barrier $D = 120\%$ implies an upper bound of 2.70%, which is more than twice as expensive as the model price 1.10% of the barrier option and its static strike-spread hedge. In the case of non-zero drift the upper bound differs even further from the model price of the option (see Neuberger and Hodges [74] for an extension of the model-independent bound to the non-zero drift case).

This little example illustrates that the completely model-independent approach of Brown, Hobson and Rogers [19] addresses the theoretically interesting question of the potential range of prices over all martingale models with the same terminal distribution, but that the resulting price bounds lie outside the usual bid-offer interval of barrier options in the OTC (*over the counter*) market. A restriction of the considered martingale processes to a class of models that resembles usual statistical properties of stock prices (like e.g. mean reversion of volatility or typical shapes and dynamics of the implied volatility surface) would certainly lead to more meaningful and much tighter bounds. But a computation of such bounds exploiting more market information does not seem to be feasible in the context of the Brown–Hobson–Rogers approach.

2.4 Towards a Practical Static Superhedge

In summary the static hedging approaches presented in the previous section have significant drawbacks which limit their applicability in practice. While the strike spread approach assumes the tradability of options with uncommon strikes and is derived under very restrictive model assumptions, the calendar spread type hedges are based on an unrealistically large number of hedge instruments and lead to instabilities as well as potentially huge hedge errors. Such hedging losses can also be infered from static mean square hedges, which by definition only weakly penalize low probability events.

Moreover, as shown in Nalholm and Poulsen [72], all these static hedges are very sensitive with respect to model risk. This is not surprising since hedge portfolios consisting of multiple standard options depend upon the curvature of the model volatility surface and how it evolves over time. Even if the model for the derivation of the hedge matches the market prices of traded options at time $t = 0$, the prices postulated by the model at the time of the barrier hit (where the static strategy specified in Definition 2.2 assumes a liquidation of the hedge) might differ significantly from observed market prices.

This conceptual problem of static hedging strategies was one of the reasons for the development of robust model-independent static superhedges. But unfortunately these approaches neglect a lot of available market information and consequently lead to price bounds which are too conservative from a practical point of view. This dilemma of model-dependent hedges with the risk of large losses on the one hand, and model-independent hedges with too wide price bounds on the other hand, summarizes the current state of the static hedging literature.

To obtain more practical static hedges which prevent losses and are robust with respect to model misspecifications, we introduce a new static super-replication approach based on optimization. The proposed approach was developed in a series of papers beginning with the initial work of Giese and Maruhn [45] and followed by papers of Maruhn and Sachs [70], [69], Maruhn [67] as well as Leibfritz and Maruhn [64]. The main idea is to identify the cheapest static hedge portfolio which super-replicates the target option in an infinite number of prespecified future market scenarios including changes of the volatility surface. This allows us to design static trading strategies that are robust against typical (not arbitrary) changes of market parameters and the dynamics of the underlying stock price. In this sense the new static hedge is located between the existing fragile model-dependent and the conservative model-independent approaches, and contains as much robustness as is needed in practice. As we will see later, the necessary robustness can often be obtained with surprisingly low cost.

In the following we will summarize the main steps for the derivation of the hedge. The starting point in Chapter 3 is the stochastic optimization problem characterizing (non-robust) cost-optimal static super-replication strategies in general financial market models. For the case of barrier options, this optimization problem has a lot of structure which can be exploited for a detailed analysis and the numerical solution of the problem. As a first step, we apply arguments from convex analysis in order to prove the existence of an optimal hedging strategy consisting of a set of traded standard options. In contrast to several other approaches we do not assume the availability of options with specific strikes and maturities – hence one cannot expect an analytic solution of the static hedging problem. However, discretizing the problem by a sample average approach (see for example Shapiro [80]) leads to a large scale linear programming problem which can be solved by standard solvers like CPLEX. As it turns out, the computed optimal static hedge portfolio for an up-and-out call in the Black–Scholes as well as Heston's stochastic volatility model clearly outperforms the approaches of Carr, Ellis and Gupta [24], Derman, Ergener and Kani [27] and Fink [38]. In addition to super-replicating the payoff of the barrier option, the identified portfolio fits the delta and vega of the target option closely, it only consists of a handful of liquidly exchange traded options and its additional premium is much smaller than half of the typical bid-offer spread of barrier options in the OTC market. Moreover, it is possible to eliminate the stochastic nature of the hedging problem by transforming it to a deterministic linear semi-infinite optimization problem which drastically speeds up numerical computations from several minutes or even hours to a few seconds.

An analysis of the hedge error surface along the barrier shows that the good performance of the new static hedge portfolio stems from an implicit regularization of the hedging problem. The strike and calendar spread approaches as well as their combination use calls to exactly fit the barrier option payoff including the discontinuity illustrated in Figure 2.1. But the payoff of a portfolio of calls as well as its valuc ovcr time is a continuous function – hence trying to exactly match the discontinuity by calls is a numerically ill-posed problem which must lead to huge hedge errors close to the discontinuity. In contrast to this, the super-replication approach approximates the discontinuous barrier option payoff (without enforcing it) by the best continuous payoff attainable through a static hedge portfolio of standard calls.

The study of the static hedging problem further raises questions regarding the structure of the dual optimization problem. In the case of dynamic trading strategies the dual of the super-replication problem is well known to be given by the maximization of the expected discounted target option payoff over the set of all equivalent martingale measures (see e.g. [30]). However, if we only consider static trading strategies (which form a finite dimensional subspace of the set of dynamic strategies) the dual problem is not known so far. But as it turns out in Chapter 4 the equivalent deterministic semi-infinite formulation can be exploited to also derive the dual problem in the static case. Surprisingly, this problem has a very intuitive interpretation. In analogy to the dynamic case, the dual problem maximizes the expected discounted barrier option payoff over a space of probability measures. But in the static setting the feasible probability measures are those measures which are consistent with the observed prices of the call options in the hedge portfolio at time $t = 0$. This theoretically shows that one of the key points in the derivation of successful static hedging strategies for barrier options is the ability of the considered model to fit the market prices of calls. The feasible measures do not have to satisfy any dynamic requirements like a martingale property, but instead only need to match the market in the sense of a static snapshot. Moreover, we are able to prove the existence of an optimal price consistent measure solving the dual problem which is the weighted sum of at most $n + 1$ Dirac-measures, where n denotes the number of standard options in the portfolio.

Although these results are very promising, the identified hedge portfolios are still non-robust in the sense that they are derived in a particular model with fixed model parameters. Consequently the hedge can still lead to huge losses if the market implied volatility surface at the time of a barrier hit differs from the surface postulated by the model. As mentioned before, this also applies to the calendar and strike spread approaches considered by Carr, Ellis and Gupta, Derman, Ergener and Kani as well as Fink. But in contrast to the latter methods, the optimization setup considered in this work will allow us in Chapter 5 to naturally incorporate additional robustness into the design of the static hedging strategy. This is achieved by a robust formulation of the hedging problem in the sense of a worst case design which guarantees the super-replication property for a pre-specified set of model parameters (corresponding to a variety of volatility surface scenarios).

Robust optimization techniques have been an active area of research in the optimization community during the last few years. For example the solution of robustified linear and quadratic programming problems can be carried out efficiently by modern optimization methods like conic or semidefinite programming (see e.g. [9], [47], [85]). Due to the success of the robust optimization framework, these ideas have also been applied to financial market problems. As one example the authors in [47] study the robust counterpart of a portfolio selection problem to make optimal portfolios less sensitive with respect to perturbations of the market data.

However, in our case the robust formulation of the static hedging problem leads to a semi-infinite optimization problem which cannot be transformed to conic or semidefinite programs such that we have to solve the problem solely relying on the theory of semi-infinite optimization. Combining arguments from this theory with Mizohata's uniqueness Theorem [71] for parabolic partial differential equations, we can prove the existence of a robust super-replication strategy in general stochastic volatility models. In addition we derive the convergence of suitable semi-infinite optimization algorithms to a solution of the robust hedging problem in the continuous time financial market model. The proposed method successively solves linear and nonlinear optimization problems and allows a very intuitive financial interpretation. First the method computes a static hedging strategy super-replicating the barrier option for a given finite set of future scenarios. But among the remaining scenarios losses may still occur. Correspondingly, we identify the worst case scenarios as solutions of appropriate nonlinear optimization problems and add these scenarios to the set of considered future outcomes. Repeating this procedure recursively leads to a sequence of hedge portfolios whose degree of robustness improves during the iteration process and finally converges to a robust static hedging strategy. Furthermore, as the numerical results for the Black–Scholes and Heston model show, the resulting robust strategies are still only marginally more expensive than the barrier option itself. This illustrates that, although model parameter uncertainty cannot be neglected, it can be taken into account with surprisingly low cost.

As mentioned before, model parameter uncertainty can also be interpreted as the risk of a changing volatility surface - including changes of the skew. The presented robust static hedging concept is the first approach in the literature which rigorously quantifies this skew risk for the case of barrier options and even allows to eliminate it. Furthermore, the associated super-replicating strategies also have a practical implication for dynamic hedge portfolios, because they provide an upper bound for the skew risk of dynamically adjusted strategies. Combined with a lower bound resulting from an analogous sub-replication problem this leads to a robust price interval for the barrier option which is still smaller than the bid-offer spread of OTC prices. This indicates that the robust optimization approach captures and efficiently exploits the structure of the volatility surface.

An additional analysis in Chapter 6 shows that although the derived static hedge portfolios are robust with respect to volatility shocks and changes of the skew, they are still very sensitive to liquidation delays or stock price jumps at the time of a barrier hit. While this equally well applies to all other static hedging approaches in the literature, the semi-infinite programming framework also allows to robustify the super-replication strategies against this source of uncertainty. The necessary modification of the optimization problem increases the dimension of the semi-infinite parameter space by one, but based on the maximum principle for parabolic partial differential equations the problem complexity can be reduced again resulting in a much simpler problem which can be interpreted as moving the barrier. As the latter is a common market practice for the dynamic hedging of barrier options, this underlines the practical relevance of the proposed approach.

Although most of the results are presented for the interesting case of an up-and-out call with discontinuous payoff profile, the theoretical and numerical results are not limited to this special class of options. Instead the static hedging framework transfers in analogy to all kinds of up/down-and-out/in barrier contracts, double knock-out barriers, options with non-zero rebates and non-constant or even discrete barriers. The various types of robustification mentioned above then can be viewed as components which may but need not be incorporated in the design of a static hedge for the barrier option under consideration.

The intuitiveness of the approach allows to even further generalize it to also take model errors into account. As an example, Chapter 7 incorporates possible deviations from model option prices to real world prices into the design of the hedging strategy. This additional robustness results in a linear semi-infinite optimization problem with a typically 15 to 30-dimensional parameter space of the semi-infinite constraints. Clearly, the high dimension of this parameter space makes the numerical solution of the problem very hard. However, by using equivalent transformations, we can reduce the dimension of the parameter space to six. The price for this reduction is a nonlinearity in the form of second order cone constraints. Employing suitable semi-infinite optimization results, we can prove convergence of an iterative method successively solving second order cone programs and nonlinear programming problems to compute the worst case constraint violation for each iterate. In addition we are able to reformulate this iterative procedure as a successive solution of semidefinite programming problems and nonlinear semidefinite programs (NSDPs). As a byproduct we obtain the result that the minimization of the minimal eigenvalue of a matrix function is equivalent to an NSDP. In particular our derivations show, that NSDPs naturally arise as subproblems of robustified linear semi-infinite programming problems to compute the constraint violation of the iterates.

Finally, Chapter 8 analyzes the empirical performance of the robust static superreplication portfolio based on a seven year dataset. On each day of the data series we set up a static hedge portfolio for an up-and-out call and a down-and-out put with one year maturity issued on the same day. Once the barrier is hit or the options expire, we record the hedge performance and gather the data in the form of detailed hedge error statistics. As expected the robustness of the static superhedging strategy against changes of the implied volatility surface and in particular changes of the skew is also confirmed empirically. A comparison with the static strike-spread hedge and a dynamic local-volatility hedge reveals that the new robust static superhedge has the smallest hedge error dispersion of all considered hedging strategies.

3 An Optimization Approach to Static Super-Replication

In this chapter we introduce the optimization based static hedging concept developed by Giese and Maruhn in [45] which is the starting point for the derivations throughout the book. Based on a given set of tradable standard options, Section 3.1 introduces the problem of super-replicating the payoff of a knock-out barrier option by means of static trading strategies in a general continuous or discrete time financial market model. After proving the existence of solutions of the resulting optimization problem, we briefly discuss the numerical solution in Section 3.2. We conclude the chapter in Section 3.3 with a detailed numerical example and a comparison with other static hedging approaches in the literature.

3.1 Cost-Optimal Static Super-Replication

Motivated by the potentially large losses of the static strike spread, calendar spread or mean square hedge approaches (see Section 2.3 for a review), we aim at deriving a hedge based on a one-sided error measure which explicitly prevents such losses. To be more precise we insist on the value of the hedge portfolio to exceed or be at least equal to the target barrier option in every possible state of the market. Some of these so-called "super-replicating strategies" may be easily found, but often they are quite expensive as it is the case for the model-independent superhedge of Brown, Hobson and Rogers [19] (see Subsection 2.3.4).

In contrast to the latter completely model-independent approach our goal in this section is to find the cheapest super-replicating hedge[1] in the set of static trading strategies, given a prespecified financial market model. Hence we combine the idea of super-replication, first introduced by El Karoui and Quenez in [30] in the context of dynamic hedges, with the idea of statically hedging barrier options by means of the knock-out trading strategies as given in Definition 2.2. Based on the general financial market model stated in Assumption 2.1 this idea can be expressed in the following definition.

Definition 3.1. Assume that $\mathcal{M}$ is a financial market model satisfying Assumption 2.1. Consider a knock-out barrier option with maturity T, payoff C_{KO} and the corresponding set of static knock-out trading strategies $\mathcal{S}_{\mathrm{KO}}$ as described in Definition 2.2. For a strategy $\phi = (\phi_t^0, \dots, \phi_t^n)_{t \in I} \in \mathcal{S}_{\mathrm{KO}}$ let $\Pi_t(\phi) = \phi_t^0 B_t + \sum_{i=1}^n \phi_t^i C_t^i$ denote the

[1] From a buyer's perspective, one can in analogy define the most expensive sub-replicating strategy (see Section 6.1).

value of the hedge portfolio at time $t \in I$. Then a "cost-optimal static super-replicating knock-out strategy" is defined as a solution of the stochastic optimization problem

$$\begin{aligned} &\min_{\phi \in \mathcal{S}_{\mathrm{KO}}} \ \Pi_0(\phi) \\ &\text{s.t.} \quad \Pi_T(\phi) \geq C_{\mathrm{KO}} \ \text{(a.s.)}. \end{aligned} \tag{3.1}$$

The set of static super-replicating knock-out trading strategies shall be denoted by

$$\mathcal{S}\mathcal{R}_{\mathrm{KO}} := \{\phi \in \mathcal{S}_{\mathrm{KO}} \ : \ \Pi_T(\phi) \geq C_{\mathrm{KO}} \ \text{(a.s.)}\}.$$

In order to illustrate the formal structure of optimization problem (3.1), we consider the following example of hedging an up-and-out call in Heston's stochastic volatility model which we will refer to in later sections.

Example 3.2. *(Heston's stochastic volatility model)*
For a given equivalent martingale measure Q, the model equations of Heston's stochastic volatility model are given by

$$dS_t = (r - \delta) S_t dt + \sqrt{Y_t} S_t dW_t^1 \quad \text{(Stock)} \tag{3.2}$$

$$dY_t = \kappa(\theta - Y_t)dt + \xi \sqrt{Y_t} dW_t^2 \quad \text{(Variance)}$$

$$dB_t = r B_t dt \quad \text{(Bond)}, \tag{3.3}$$

where r is the constant riskfree rate and δ is the constant dividend yield. The variance Y of S is modeled by a square-root process, where κ denotes the mean reversion rate, θ is the long-run mean of the variance and ξ the volatility of volatility. W^1 and W^2 are standard Brownian motions with correlation coefficient ρ. In terms of Assumption 2.1 we have $I = [0, T]$ for some $T > 0$ and the filtration is given by $\left((\mathcal{F}_t^W)^*\right)_{t \in I}$, i.e. the augmented Filtration generated by $W = (W^1, W^2)$.

Our goal is to hedge an up-and-out-call $C_{\mathrm{KO}} := (S_T - K)^+ \cdot 1_{\{\max_{t \in I} S_t < D\}}$ with maturity T, barrier $D > S_0$ and strike $K > 0$ by a static portfolio of standard calls and binary calls denoted by $C^1, \ldots, C^n$ with potentially varying maturities $T_i \leq T$ and the riskless bond B. For $t > T_i$ the arbitrage-free value processes of the options $C^1, \ldots, C^n$ are extended pathwise by setting

$$C_t^i = C_{T_i}^i \frac{B_t}{B_{T_i}}, \quad T_i \leq t \leq T, \tag{3.4}$$

which simulates shifting the payoff of C^i to the bond position as soon as option C^i expires. Thus the market model $\mathcal{M}$ consists of Heston's stochastic volatility model under the equivalent martingale measure Q, where the financial instruments allowed for trade are S, B described by model equations (3.2), (3.3) and $C^1, \ldots, C^n$ with arbitrage-free value processes extended by (3.4). Hence optimization problem (3.1) is given by

$$\begin{aligned}
&\min_{\phi \in \mathcal{S}_{\mathrm{KO}}} \Pi_0(\phi) \\
\text{s.t. } \Pi_T(\phi) &\geq C_{\mathrm{KO}} \text{ (a.s.)} \\
C_t^i &= B_t E_Q\left(C_{T_i}^i / B_{T_i} | \mathcal{F}_t\right), \quad 0 \leq t \leq T_i \\
C_t^i &= C_{T_i}^i \frac{B_t}{B_{T_i}}, \quad T_i < t \leq T \qquad (3.5)\\
dS_t &= (r - \delta) S_t dt + \sqrt{Y_t} S_t dW_t^1 \\
dY_t &= \kappa(\theta - Y_t) dt + \xi \sqrt{Y_t} dW_t^2 \\
dB_t &= r B_t dt.
\end{aligned}$$

Thus the cost-optimal static super-replicating strategy is a solution of the stochastic optimization problem (3.1). The question remains whether such a solution really exists, that means whether the infimum of Π_0 is attained on the feasible set $\mathcal{S}\mathcal{R}_{\mathrm{KO}}$. As it turns out, this question can be answered in a very general setting. In Appendix A we prove the existence of cost-optimal strategies under mild conditions and independently of the structure of the financial market model under consideration and the target claim that shall be super-replicated. The following theorem states the existence result in the case of super-replicating a knock-out option by means of static knock-out strategies.

Theorem 3.3. *Let $\mathcal{M}$ be a financial market model as described in Assumption 2.1. Further let C_{KO} and $\mathcal{S}_{\mathrm{KO}}$ be given as stated in Definition 2.2. If $\mathcal{S}\mathcal{R}_{\mathrm{KO}} \neq \emptyset$ and $\mathcal{M}$ is arbitrage-free in the set of trading strategies $\mathcal{S}_{\mathrm{KO}}$, then a cost-optimal static super-replicating knock-out strategy exists, i.e. the stochastic optimization problem (3.1) has a solution.*

If furthermore an exact hedging strategy exists in $\mathcal{S}_{\mathrm{KO}}$, i.e. $\Pi_T(\phi) = C_{\mathrm{KO}}$ (a.s.) for some $\phi \in \mathcal{S}_{\mathrm{KO}}$, then every cost-optimal static super-replicating knock-out strategy is also exact and the price of the hedge portfolio at $t = 0$ coincides with the price of the exact static hedge. In particular the cost-optimal static super-replicating strategy equals the exact hedge in $\mathcal{S}_{\mathrm{KO}}$, if the latter is unique.

Proof. By definition the set of static knock-out trading strategies $\mathcal{S}_{\mathrm{KO}}$ is isomorphic to $\mathbb{R}^{n+1}$. Thus we can apply the general existence Theorem A.4 in Appendix A and immediately obtain the existence of a solution. The second part of the theorem is a simple application of Theorem A.5 in the Appendix. □

Note that any market model, which is arbitrage-free in the set of dynamic trading strategies satisfying the usual regularity assumptions, is also arbitrage-free in the set of static knock-out trading strategies $\mathcal{S}_{\mathrm{KO}}$ given in Definition 2.2. Hence Theorem 3.3

can be applied to a broad range of financial market models (Black–Scholes model, Heston's stochastic volatility model, local volatility models etc.). In those models the theorem guarantees the existence of a cost-optimal static super-replicating strategy in case there exists at least one static strategy in $\mathcal{S}_{\mathrm{KO}}$ super-replicating the payoff of the barrier option. The latter assumption can usually be fulfilled by including appropriate standard options in the set of tradable instruments $C^1, \ldots, C^n$. For example the up-and-out-call $C_{\mathrm{KO}} := (S_T - K)^+ \cdot 1_{\{\max_{t \in I} S_t < D\}}$ is super-replicated by a standard call with strike K and maturity T.

In addition to the existence, Theorem 3.3 implies that a cost-optimal static super-replicating strategy is exact if an exact static hedging strategy exists. Thus we can conclude, that in the Black–Scholes model with zero cost of carry, the exact static hedging strategy as derived by Bowie and Carr [16] (see Subsection 2.3.1) is a solution of optimization problem (3.1) if the standard options identified in [16] are among the tradable options $C^1, \ldots, C^n$.

However, the results of Bowie and Carr [16] as well as other exact static hedging approaches are derived in simple models and rely on the crucial assumption that specific hedge instruments can be traded in the market. If these assumptions are dropped, it is not clear how to set up a static hedge portfolio. This in turn shows the generality of the concept of cost-optimal static super-replication where we can simply include the hedge instruments in the portfolio that are truly tradable in the market to derive the best attainable static hedge portfolio. To compute the cost-optimal static super-replication portfolio, that is to solve problem (3.1), appropriate optimization methods have to be applied. This is the focus of the next section.

3.2 Numerical Implementation

In general an analytic computation of the cost-optimal static super-replicating strategy derived in the last section will not be possible such that we have to numerically solve problem (3.1). In Chapter 4 we will show that this stochastic optimization problem is in fact equivalent to a semi-infinite optimization problem allowing us to employ advanced optimization techniques to solve the problem. But for the moment we first pursue a more natural discretization which can be obtained by replacing the infinite number of constraints $\Pi_T(\phi)(\omega) \geq C_{\mathrm{KO}}(\omega)$ for all $\omega \in \Omega \setminus N$, where $N \in \mathcal{F}$ denotes a set of measure zero, with $M \in \mathbb{N}$ samples $\omega_k \in \Omega$. The resulting discretized problem is given by

$$\begin{aligned} &\min_{\phi \in \mathcal{S}_{\mathrm{KO}}} \Pi_0(\phi) \\ &\text{s.t. } \Pi_T(\phi)(\omega_k) \geq C_{\mathrm{KO}}(\omega_k), \; k = 1, \ldots, M. \end{aligned} \tag{3.6}$$

A feasible point of this problem will guarantee the super-replication property for the M considered scenarios. Hence, by intuition, M has to be chosen quite large to sufficiently approximate the original optimization problem (3.1) which in turn increases the computational effort to solve the problem. But note that by Definition 2.2 the set of static knock-out trading strategies $\mathcal{S}_{\mathrm{KO}}$ is isomorphic to $\mathbb{R}^{n+1}$. In particular we have

$$\begin{aligned}\Pi_t(\phi) &= \alpha_0 B_t + \frac{\sum_{i=1}^n \alpha_i C^i_{\tau\wedge T}}{B_{\tau\wedge T}} 1_{\{\tau\le t\}} B_t + \sum_{i=1}^n \alpha_i 1_{\{\tau>t\}} C^i_t \\ &= \alpha_0 B_t + \frac{B_t}{B_{\tau\wedge t}} \sum_{i=1}^n \alpha_i C^i_{\tau\wedge t} \end{aligned} \tag{3.7}$$

such that $\Pi_t(\phi) = \Pi_t(\alpha)$, where $\alpha = (\alpha_0, \ldots, \alpha_n)^{\mathrm{T}} \in \mathbb{R}^{n+1}$ denotes the unique vector corresponding to the trading strategy $\phi \in \mathcal{S}_{\mathrm{KO}}$. Furthermore, by equation (3.7), the mapping $\Pi_t(\cdot)(\omega) : \mathbb{R}^{n+1} \to \mathbb{R}$ is linear $\forall\, t \in [0, T]$, $\forall\, \omega \in \Omega$. This implies that the discretized optimization problem (3.6) is a linear programming problem of the form

$$\min_{\alpha\in\mathbb{R}^{n+1}} c^{\mathrm{T}}\alpha \quad \text{s.t.} \quad A\alpha \ge b, \tag{3.8}$$

where $A \in \mathbb{R}^{M\times(n+1)}$, $b \in \mathbb{R}^M$ and $c \in \mathbb{R}^{n+1}$ are chosen appropriately. Due to (3.7) and (3.6) we can determine A, b and c by computing the stopping time τ of the first barrier event, the value of the European options C^i at the stopping time as well as the payoff of the barrier option C_{KO} at maturity. Further note that the problem data A, b and c is independent of the variable α such that we can separate the generation of the data from the solution of the optimization problem (3.8). Regarding the implementation this means that we can store the generated option prices at the stopping time or at T for each sample path and then solve the resulting large scale linear programming problem. These findings are summarized in Algorithm 1 which gives a step-by-step procedure for the Monte Carlo-based solution of the super-replication problem (3.1).

While the determination of the stopping time $\tau(\omega_k)$ and the payoff of the barrier option $C_{\mathrm{KO}}(\omega_k)$ is a straight forward task, the computation of the value of the European options C^i in case the first barrier event occurs before T usually requires evaluating a conditional expectation of the form $B_t E_Q(\frac{1}{B_{T_i}} C^i_{T_i} | \mathcal{F}_t)$, see e.g. Example 3.2. For this purpose we can either use a Monte Carlo approximation, a numerical method to solve the pricing partial differential equation, or a closed form solution if available. Although the Monte Carlo approach is possible, it is not recommended as this method has to be employed for each of the M sample paths for which the first barrier event occurs before T, hence requiring a lot of computing time.

Fortunately, in case of Heston's stochastic volatility model, we can make use of the following closed form solution for the value of a standard call C^i with maturity T_i and strike K_i at time $t < T_i$ derived by Heston in [50]

$$C^i_t = S_t e^{-\delta(T_i - t)} P_1 - K_i e^{-r(T_i - t)} P_2$$

$$P_j = \frac{1}{2} + \frac{1}{\pi}\int_0^\infty \mathrm{Re}\left(\frac{e^{-\iota\psi \ln K_i} f_j(x, Y_t, t, \psi)}{\iota\psi}\right) d\psi, \ j = 1, 2$$

Algorithm 1: Monte Carlo-Based Solution Method

Input: Number of samples $M \in \mathbb{N}$.

Main Algorithm:

for $k = 1, \dots, M$ **do**

Simulate one sample path $(S_t(\omega_k))_t$ of the underlying.

Determine whether and when the first barrier event occurs for the given sample, i.e. compute the value of the stopping time $\tau(\omega_k)$.

if $\tau(\omega_k) \leq T$ **then**

Calculate the value of the European options C^i at the stopping time and accumulate this value until T. Store the result $C^i_{\tau(\omega_k)} B_T / B_{\tau(\omega_k)}$ as the matrix component $a_{k,i+1}$, $i = 1, \dots, n$.

else

Determine the value of the European options C^i at T and store this price as $a_{k,i+1} = C^i_T(\omega_k)$, $i = 1, \dots, n$.

end

Calculate the payoff of the barrier option $C_{\mathrm{KO}}(\omega_k)$ for the given sample path, set $b_k = C_{\mathrm{KO}}(\omega_k)$ and $a_{k,1} = B_T$.

end

Set $c_1 = B_0$, $c_{i+1} = C^i_0$, $i = 1, \dots, n$, and solve the (large scale) linear programming problem (3.8) with $A = (a_{k,i})_{k,i} \in \mathbb{R}^{M \times (n+1)}$ and vectors $b = (b_1, \dots, b_M)^{\mathrm{T}}$, $c = (c_1, \dots, c_{n+1})^{\mathrm{T}}$.

$$
\begin{aligned}
f_j(x, Y_t, t, \psi) &= \exp(D_j(T_i - t, \psi) + E_j(T_i - t, \psi)Y_t + \iota\psi x) \\
D_j(\eta, \psi) &= \iota(r - \delta)\eta\psi + \frac{a}{\xi^2}\left((b_j - \iota\rho\xi\psi + d_j)\eta - 2\ln\left(\frac{1 - g_j e^{d_j \eta}}{1 - g_j}\right)\right) \\
E_j(\eta, \psi) &= \frac{b_j - \iota\rho\xi\psi + d_j}{\xi^2}\left(\frac{1 - e^{d_j \eta}}{1 - g_j e^{d_j \eta}}\right) \\
g_j &= \frac{b_j - \iota\rho\xi\psi + d_j}{b_j - \iota\rho\xi\psi - d_j} \\
d_j &= \sqrt{(\iota\rho\xi\psi - b_j)^2 - \xi^2(2\iota u_j \psi - \psi^2)}
\end{aligned}
$$

$$
x = \ln(S_t),\ u_1 = \frac{1}{2},\ u_2 = -\frac{1}{2},\ a = \kappa\theta,\ b_1 = \kappa - \rho\xi,\ b_2 = \kappa.
$$

Here ι denotes the imaginary unit and S_t, Y_t are the values of the stock and variance at time t. The remaining variables are given by the parameters of Heston's stochastic volatility model as introduced in Example 3.2.

Once the conditional expectations are calculated and thus the data of the large scale linear programming problem is set up, we can compute its solution by appropriate linear programming solvers. If we neglect possible discretization errors in the simulation of the stock price, we can additionally expect the solution of the linear program (3.8) to converge to the solution of the underlying stochastic optimization problem (3.1) as we increase the number of Monte Carlo simulations M. This convergence behavior can for example be verified by transforming problem (3.1) to the equivalent problem

$$\min_{\phi \in \mathcal{S}_{\text{KO}}} \Pi_0(\phi)$$
$$\text{s.t.} \quad P(\Pi_T(\phi) < C_{\text{KO}}) = 0.$$

Estimating the probability by a Monte Carlo approximation with $M \in \mathbb{N}$ samples $\omega_k \in \Omega$, we obtain the discretization

$$\min_{\phi \in \mathcal{S}_{\text{KO}}} \Pi_0(\phi)$$
$$\text{s.t.} \quad \frac{1}{M} \sum_{k=1}^{M} 1_{\{\Pi_T(\phi) < C_{\text{KO}}\}}(\omega_k) = 0.$$

As all indicators in the sum of the constraint are greater or equal to zero, this optimization problem is in fact equivalent to the large scale linear programming problem (3.6). Hence (3.6) can be interpreted as a sample average approximation (see e.g. Shapiro [80]) of the original stochastic optimization problem (3.1) such that the usual convergence results of the sample average theory apply. As an alternative, the convergence of the solutions can also be derived by epiconvergence results as for example presented by Pennanen and Koivu [76].

3.3 Results and Comparison to other Hedging Approaches

In this section we provide some examples for the cost-optimal static super-replication of barrier options. In particular we compare the performance of the newly developed hedge with the results of other methods in the existing literature which will prove the superiority of the cost-optimal static super-replication method.

3.3.1 Static Hedging in the Black–Scholes Model

Consider an up-and-out call option with strike $K = 2750$, barrier $D = 3300$, rebate $R = 0$ and a maturity of one year ($T = 1$) on an underlying asset with a current spot price of $S_0 = 2750$ which roughly corresponds to the average price of the EURO STOXX 50 in September 2004. First, we try to hedge the barrier option in the classic Black–Scholes model with a risk-free interest rate $r = 5.5\%$, a dividend yield $\delta = 2.5\%$ and a constant volatility of 20%.

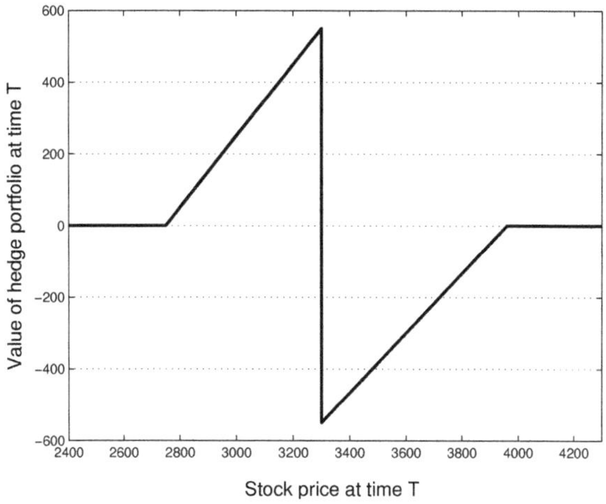

Figure 3.1: Graphical illustration of the Bowie Carr hedge portfolio

If the interest rate and dividend yield would be equal, one could perfectly replicate the barrier option in the Black–Scholes model based on the strike spread approach (see Subsection 2.3.1 for details). The corresponding hedge portfolio consists of 1.0 standard calls with strike 2750, $-\frac{1}{6}$ standard calls with strike 3300, $-\frac{5}{6}$ standard calls with strike 3960 and -1100 binary calls with strike 3300, each with one year maturity, as shown by Bowie and Carr in [16] (hereafter BC). The final payoff of this so-called strike spread hedge[2] is illustrated in Figure 3.1.

However, here we consider the practically more interesting case of nonzero drift which would theoretically require an infinite number of standard options to match the nonlinear adjusted payoff function (2.4). Since we have to limit ourselves to a finite number of options, we consider the strike spread sketched in Figure 3.1 as an approximation of the exact adjusted payoff (2.4). But even after this necessary approximation it is difficult to set up the BC portfolio in practice, because one year calls with strikes above 140% of the current underlying spot as well as binary calls are rarely traded in the market. The question is how the cost-optimal static superhedge performs in comparison to the BC portfolio in a non-zero drift environment and only making use of liquidly traded standard calls.

As another benchmark we consider the calendar spread approach (see Subsection 2.3.2) of Derman, Ergener and Kani (hereafter DEK), who suggest in [27] to hedge the barrier option using a hedge portfolio which consists of 24 calls $C^1, \dots, C^{24}$ with strikes equal to the barrier D and varying maturities $T_i = \frac{i}{24}, i = 1, \dots, 24$, plus one

[2] It is worth mentioning that if the dividend yield is set equal to the interest rate, then the BC portfolio is correctly identified by Algorithm 1, that is the BC portfolio equals the cost-optimal static super-replicating strategy as proven in Theorem 3.3.

additional call C^{25} with strike and maturity equal to the strike and the maturity of the up-and-out call. The call C^{25} with strike $K = 2750$ replicates the payoff of the barrier option if the barrier is never hit. On the other hand, the quantities α_i of the options C^i, $i = 1, \ldots, 24$, are chosen recursively such that the total value of the hedge portfolio is exactly zero on the barrier $D = 3300$ at all half-month time points $t_j = \frac{j-1}{24}$, $j = 1, \ldots, 24$ before expiration. Note that in principle the approach of DEK allows to construct a hedge portfolio which matches the barrier option on the barrier at many more time points, however, that would require as many additional options as additional time points which is problematic since the ability to trade all of the 24 options with strike 3300 in the market is already questionable.

Instead of using options which are not available in the market or even unrealistically many different hedge instruments, we choose to construct a static super-replication portfolio only consisting of EURO STOXX 50 options which are liquidly tradable on the EUREX so that buying the obtained optimal hedging portfolio would indeed be possible. After applying Algorithm 1 first using the whole set of EURO STOXX 50 options tradable on the EUREX in September 2004, we identified the ten options listed in Table 3.1 to be most suited for hedging the up-and-out call option under consideration. Therefore, we will restrict our analysis to finding the cost optimal static super-replication portfolio using only the options listed in Table 3.1.

Maturity in Years	*Strike*	*Optimal Quantity*
1.00	2750	1.000
1.00	3300	−33.034
1.00	3350	40.085
1.00	3450	−8.222
1.00	3600	0.807
0.75	3300	−0.037
0.75	3400	0.046
0.75	3600	−0.253
0.50	3300	0.016
0.50	3500	−0.094

Table 3.1: Hedge instruments and optimal quantities for the cost-optimal static super-replication portfolio in the Black–Scholes model

In order to obtain the optimal portfolio weights, we apply Algorithm 1 with $M = 1\,000\,000$ sample paths and 20 000 time-steps per path[3]. The resulting quantities (rounded to the third decimal place) for the selected ten hedge instruments are presented in Table 3.1. Obviously the main part of the portfolio consists of several calls with maturity $T = 1$, but different strikes (*strike-spread*). The value of this part of

[3]To correct for the bias of the discrete simulation the Brownian bridge correction algorithm as described in [3] was applied.

the portfolio at time T is illustrated in Figure 3.2 for a range of stock prices. Apparently the shape of the curve is similar to the corresponding graph of the BC portfolio (see Figure 3.1). Looking at the portfolio positions in Table 3.1, one can quickly observe that the second and third call approximate the Bowie Carr binary position by a call spread. The rest of the calls with maturity $T = 1$ seem to be a nontrivial approximation of the BC portfolio for the non-zero drift case based on calls traded in the market. Furthermore, the calls in the portfolio expiring before T correct possible portfolio mismatches for various times along the barrier (*calendar spread*). Hence the super-replication approach combines calendar- and strike-spreads in an optimal way based on actually tradable instruments.

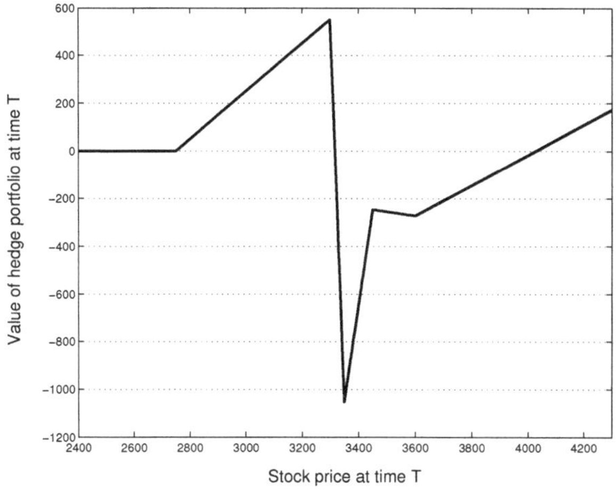

Figure 3.2: Graphical illustration of the strike spread component of the super-replication portfolio

The question remains how the identified portfolio performs in comparison to the BC- and DEK-portfolio. As a first analysis, Table 3.2 shows the prices[4] and sensitivities of the barrier option and the different static hedge portfolios. Surprisingly, the value of the cost-optimal static super-replication portfolio (1.14% of the underlying spot price) is only about 1 basis point above the fair value of the barrier option (1.13%).

Table 3.3 presents statistics of the distribution of the cash-flows generated from selling the barrier option and holding the corresponding hedge portfolio based on 1 000 000 sample paths. The BC portfolio, which would provide a perfect hedge if interest rate and dividend yield are equal, clearly fails to hedge the considered up-and-out call properly in a non-zero drift environment. In contrast to the BC portfolio, the

[4] Here, the prices of options are always expressed in percent of the underlying spot as it is common in the OTC market.

	Barrier Option	*BC*	*DEK*	*Super Replication*
Price	1.13%	1.00%	1.23%	1.14%
Vega	−0.12%	−0.11%	−0.42%	−0.10%
Delta	−0.017	−0.025	−0.018	−0.017

Table 3.2: Price and sensitivities of the barrier option and the considered hedge portfolios in the Black–Scholes model

cost-optimal static super-replication portfolio and the DEK portfolio both always produce a payout which is at least as high as the payoff of the barrier option, while being only slightly more expensive than the barrier option. Among the two, the cost-optimal static super-replication portfolio is superior by having a lower price, a lower standard deviation and a considerably smaller number of hedge instruments than the DEK portfolio. Also note that the vega of the DEK portfolio is more than three times the vega of the barrier option being hedged whereas the cost-optimal super-replication portfolio matches the delta and vega of the barrier option closely.

	BC	*DEK*	*Super Replication*
Average	−0.13%	0.10%	0.01%
Standard Deviation	0.17%	0.86%	0.28%
Minimum	−0.44%	0.00%	0.00%

Table 3.3: Hedge error statistics in the BS-model (1 000 000 simulations)

The superiority of the static super-replication approach is also graphically visible in Figure 3.3 which shows the mismatch in value on the barrier between the hedging portfolios and the barrier option for hitting times between 0.8 and 1.0 years.

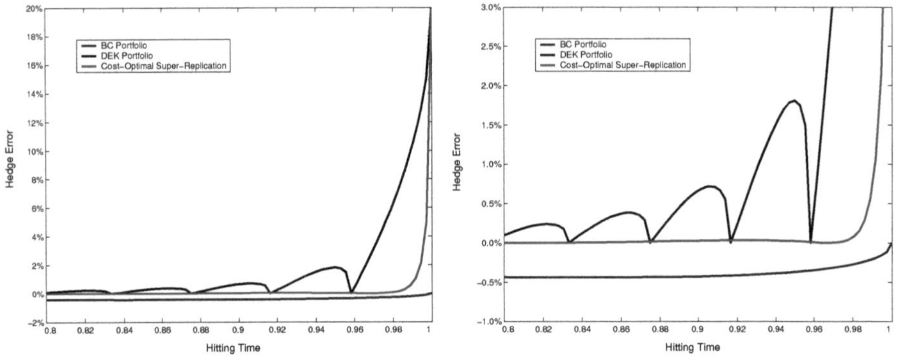

Figure 3.3: Hedge error on the barrier for various hitting times in the Black–Scholes model (right: zoom-in)

Although the BC portfolio contains a binary position and hence its value on the barrier close to T approaches zero, the non-zero drift leads to huge negative hedge errors within a large time interval. In contrast to this the graphs of the DEK- and super-replication portfolio always stay non-negative and bend upwards to the inner value of the barrier option while approaching the maturity T. But compared to the DEK-portfolio, the cost-optimal super-replication graph is much flatter on the whole time interval also implying the significantly lower cost. In fact the value function of the cost-optimal super-replication portfolio is very close to the discontinuous up-and-out call payoff. The only difference is that the value function evolves continuously in time as well as stock price and hence has to connect the knock-out value of zero to the inner value $D - K$. Thus the super-replication implicitely regularizes the hedging problem by choosing the best continuous approximation of the discontinuous barrier option payoff. In the next subsection we will analyze if these very attractive properties also transfer to a stochastic volatility environment.

3.3.2 Static Hedging in Heston's Stochastic Volatility Model

As illustrated in Toft and Xuan [83], the quality of a static hedge of barrier options is very sensitive to the volatility assumption. By analyzing the DEK calendar spread hedge in Heston's stochastic volatility model, the authors demonstrate that the effectiveness of this particular static hedging method deteriorates rapidly and considerable losses can occur when the volatility is stochastic and the volatility of volatility is large.

Toft and Xuan identified that one main reason for the poor performance of the DEK approach in a stochastic volatility environment is the tendency of the method to produce hedge portfolios with vegas that differ significantly from the vega of the barrier option being hedged (as can also be seen in Table 3.2). Consequently, a volatility shock will have a much larger impact on the value of the hedge portfolio than on the value of the target barrier option which can easily result in large hedging losses.

Motivated by these findings Fink [38] extended the DEK hedge approach by constructing a static hedge portfolio whose value matches the value of the barrier option on the barrier not only for a set of time points but also for a set of volatility states (see Subsection 2.3.2). The vega of the resulting hedge portfolio matches the vega of the barrier option quite well and the performance of the static hedge in a stochastic volatility environment is considerably improved in comparison to the DEK portfolio. However, in order to construct a static hedge which matches the barrier option's payoff profile on the barrier at n_t time points and for n_v volatility states, one needs at least $n_t \times n_v + 1$ standard options which easily accumulates to an unmanageable huge number of different hedge instruments whose tradability in the market would be questionable. Furthermore, even this improved vega neutral static hedge deteriorates if the volatility of volatility is high as shown in [38].

Consequently, we need to examine the performance of the cost-optimal static super-replication method in a stochastic volatility model with high volatility of volatility. For this reason we follow Toft and Xuan [83] as well as Fink [38] and consider Heston's stochastic volatility model (see equations (3.2), (3.3)) with a risk-free interest rate $r = 5.5\%$, a dividend yield $\delta = 2.5\%$, a start variance $Y_0 = 0.04$, a long-run mean $\theta = 0.04$, a mean reversion rate $\kappa = 1.5$, a volatility of volatility $\xi = 0.2$ and a correlation between the variance and the asset price process $\rho = -0.5$. The current price of the underlying is assumed to be 2750 and the hedging target is again a 1-year up-and-out call option with strike 100% and barrier 120% of the underlying spot.

In order to hedge this up-and-out call Fink [38] uses 254 different options and extends the standard DEK approach by constructing a static hedge portfolio whose value matches the value of the barrier option on the barrier in the sense of conditions (2.6) at 63 time points $t_j = \frac{j-1}{63}$, $j = 1, \dots, 63$ and for 4 different variance states $\bar{y}^1 = 0.040$, $\bar{y}^2 = 0.044$, $\bar{y}^3 = 0.057$ and $\bar{y}^4 = 0.067$.

In contrast to the usage of unrealistically many, not tradable options we again restrict the cost-optimal static super-replication portfolio for the target barrier option to contain only the preselected exchange traded options listed in Table 3.1. The cost-optimal static super-replication hedge is then obtained by applying Algorithm 1 with $M = 1\,000\,000$ sample paths and 20 000 time-steps. The resulting optimal quantities of the hedge instruments are presented in Table 3.4. As in the Black–Scholes case the portfolio is a combination of strike and calendar spreads including an approximation of the Bowie Carr binary position.

Maturity in Years	*Strike*	*Superhedge*	*Mean-Square Hedge*
1.00	2750	1.000	0.978
1.00	3300	−25.479	−38.127
1.00	3350	29.718	53.637
1.00	3450	−6.410	−21.623
1.00	3600	1.655	5.742
0.75	3300	0.364	0.381
0.75	3400	−0.244	0.111
0.75	3600	−0.508	−1.022
0.50	3300	0.174	0.157
0.50	3500	−0.095	−0.015

Table 3.4: Hedge instruments and optimal quantities (rounded to the third decimal place) for the cost-optimal static super-replication portfolio and the mean-square hedge in Heston's stochastic volatility model

As an additional benchmark we also consider the static mean-square hedging approach of Dupont [29] introduced in Subsection 2.3.3. In contrast to the static super-replication portfolio this particular hedging strategy aims at a minimization of the L^2-hedge error (2.7). Based on a discretization of this minimization problem in the spirit of Section 3.2, we can compute the mean square hedge by solving the resulting quadratic optimization problem. Applying this method to the same set of potential hedge instruments as in the super-replication case, we obtain the optimal portfolio weights listed in Table 3.4. Obviously, the mean-square hedge also combines strike and calendar spreads to minimize the L^2-error measure. However, compared to the superhedge the Bowie Carr binary position is approximated with much more aggressive portfolio weights. Furthermore, the call with the same strike and maturity as the barrier option is not included with a unit weight of 1.0.

The price and the sensitivities of the barrier option and those of the three hedging portfolios are given in Table 3.5, while Table 3.6 lists the statistics of the cash-flow distribution generated from selling the barrier option and holding the corresponding hedge portfolio based on 1 000 000 sample paths. Furthermore, Figures 3.4, 3.5 and 3.6 graph the resulting distribution of the cash-flows for the Fink portfolio, the cost-optimal static superhedge and the mean-square hedge portfolio.

	Barrier Option	*Fink*	*Mean-Square Hedge*	*Superhedge*
Price	1.60%	1.61%	1.56%	1.79%
Vega[5]	−0.19%	−0.20%	−0.18%	−0.19%
Delta	−0.034	−0.033	−0.034	−0.031

Table 3.5: Price and sensitivities of the barrier option and the considered hedge portfolios in Heston's stochastic volatility model

	Fink	*Mean-Square Hedge*	*Superhedge*
Average	0.01%	−0.04%	0.19%
Standard Deviation	0.30%	0.66%	0.83%
Minimum	−4.43%	−7.32%	0.00%

Table 3.6: Statistics of hedge error in Heston's model (1 000 000 simulations)

The value of the cost-optimal static super-replication portfolio (1.79% of the underlying spot price) is 19 basis points above the fair value of the barrier option (1.60%). The price difference is much larger than in the Black–Scholes case, however, it is less than half of the typical bid-ask spread for barrier options in the OTC market. Furthermore, the cost-optimal static super-replication portfolio matches the delta and the vega of the barrier option closely although this property is not required to hold in the description of the optimization problem.

The Fink portfolio also matches the delta and the vega of the barrier option, while in addition fitting the fair value of the up-and-out call almost perfectly, being only 1 basis point more expensive. However, the Fink portfolio can result in substantial hedging losses, which are a multiple of the initial value of the barrier option. Table 3.6 and Figure 3.4 reveal, that losses occur for a considerable number of scenarios including

[5] We calculate the vega in Heston's stochastic volatility model by shifting the start variance Y_0 and the mean reversion variance level θ such that the corresponding volatility quantities are shifted by an absolute size of $\pm 1\%$. Taking the average of the recalculated option prices produces a vega which is intuitively comparable to the vega in the Black–Scholes model.

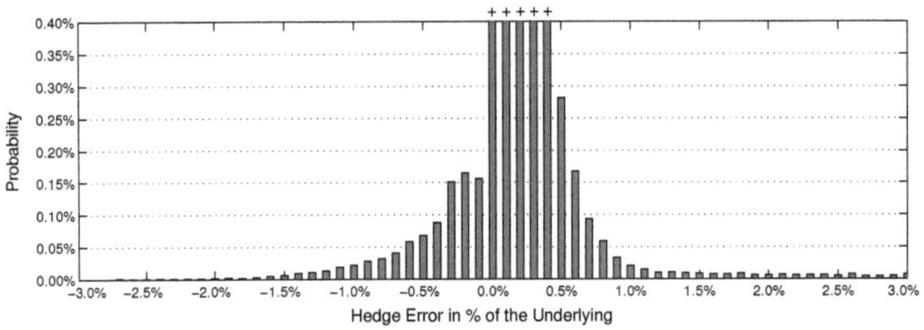

Figure 3.4: Histogram of the hedge error for Fink's portfolio

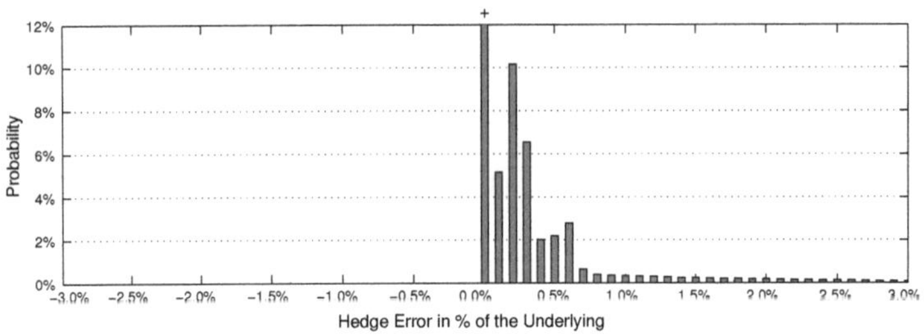

Figure 3.5: Histogram of the hedge error for the super-replication portfolio

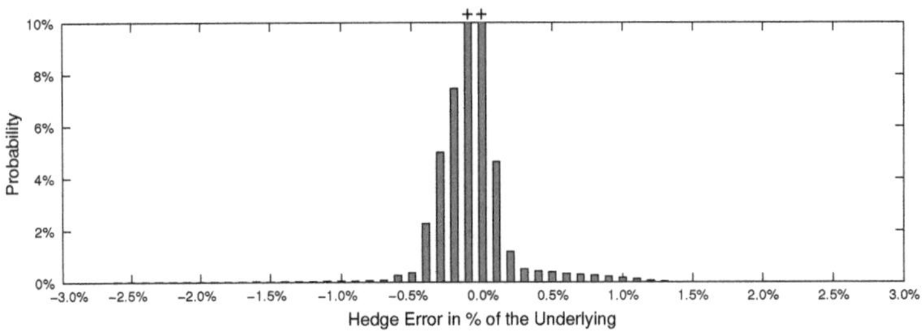

Figure 3.6: Histogram of the hedge error for the mean-square hedge portfolio

a maximum loss of -4.43%. Typically, these large hedging losses are realized when the barrier is hit near expiration. This can be seen for instance in Figure 3.7 which shows the value of Fink's hedging portfolio at the barrier for a broad range of barrier hitting times and variance realizations. Besides the largely negative values for low volatility states close to maturity T the rugged surface of the graph reveals the ill-posedness of Fink's approach which arises by trying to force the hedge error function to exactly zero. These results imply that the DEK static hedging approach[6] as well

[6] In [38] Fink shows, that his approach clearly outperforms the DEK approach in a stochastic volatility

as Fink's vega-neutral extension perform very poorly in exactly the same situations as those where the classical dynamic hedging strategies fail to hedge barrier options. This makes these hedging approaches essentially useless for barrier option traders who are mostly concerned about the worst case of hitting the barrier close to maturity.

The mean-square hedge struggles with similar problems. Even though the portfolio matches the price and sensitivities of the barrier option quite well, the hedge can lead to significant losses of up to -7.32%, which is more than 4.5 times the fair value of the barrier option. Compared to the superhedge the L^2-hedge error has a slightly smaller standard deviation (which is not surprising since the standard deviation is closely related to the L^2-error measure), but the cost for this reduction is too high if the risk of extremely large losses is taken into account. Figure 3.6 illustrates the typical shape of a hedge error distribution for a static portfolio derived with the L^2-measure. The distribution is more or less symmetric and centered at zero, with quickly decreasing probability mass at the wings. Consequently, potentially large hedging losses are possible, but they only occur with low probability.

In contrast to this, the cost-optimal static super-replication portfolio always results in a non-negative cash-flow for the considered range of barrier hitting times and variance realizations at the barrier as shown in Figure 3.8. Compared to Fink's hedge error surface, the super-replication surface looks much smoother and hence graphically illustrates the well-posedness of the proposed static hedging approach. In particular Figure 3.8 reveals that the computed optimal super-replication portfolio is the natural extension of the Black–Scholes portfolio and its corresponding hedge error surface shown in Figure 3.3. The optimized portfolio tries to keep the hedge error function as close to zero as possible, but shortly before maturity T the surface has to bend upwards to the inner value of the barrier option. Hence the hedging portfolio offers a comfortable safety cushion in the form of large positive liquidation values especially for the critical hitting times close to maturity. This property as well as the fact that the portfolio consists of only a handful of exchange tradable options makes the cost-optimal static super-replication method appealing to option traders and of practical relevance.

In summary the presented results show that the new hedging method outperforms the other static hedging approaches in the literature. The superiority of the method was demonstrated in a constant volatility model as well as in a stochastic volatility model with high volatility of volatility. The clear advantages of the method are the usage of hedge instruments which are truly available in the market and the avoidance of hedging losses by super-replicating the payoff of the barrier option in any state of the postulated market model. In addition, the cost-optimal static super-replication portfolio matches the delta and vega of the target barrier option very closely and is only slightly more expensive than the barrier option itself. The only drawback so far is the computation time of Algorithm 1 which is majorly due to the huge number of Monte Carlo simulations needed to obtain sufficiently accurate results. In the next chapter we will see that it is possible to transform the stochastic optimization problem

setting. For that reason we restrict our comparison to Fink's portfolio.

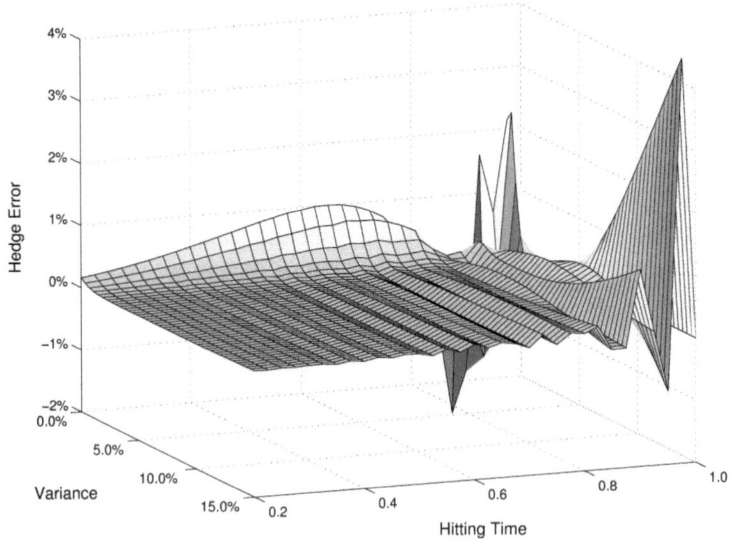

Figure 3.7: Hedge error on the barrier for Fink's hedge portfolio

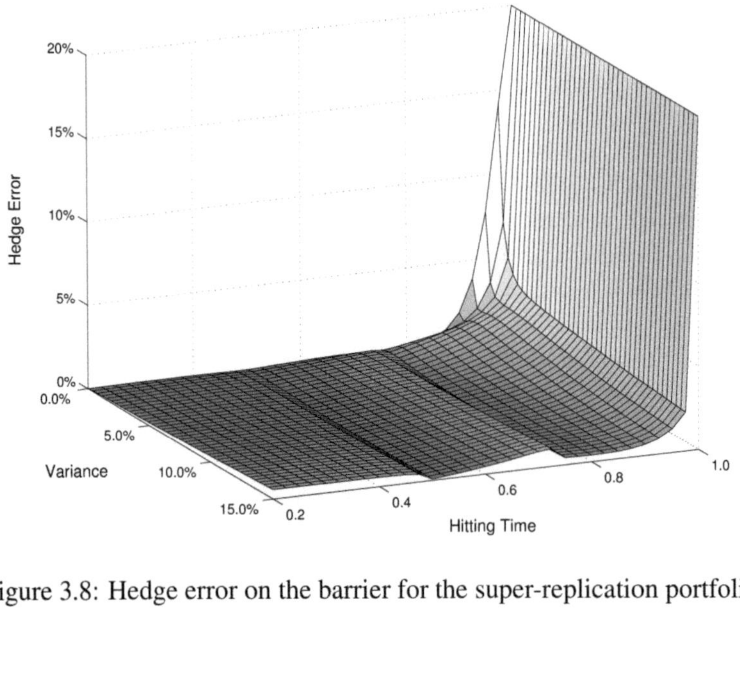

Figure 3.8: Hedge error on the barrier for the super-replication portfolio

to a purely deterministic problem which drastically speeds up computation time from several minutes to a few seconds.

4 Reformulation as a Semi-Infinite Problem

In this chapter we prove that the super-replication problem can be equivalently reformulated as a deterministic semi-infinite optimization problem. Based on an analysis of the support of suitable distributions, Section 4.1 shows that it is possible to completely eliminate the stochastic nature from optimization problem (3.1). As we will see in Chapter 5, this semi-infinite equivalence allows to apply purely deterministic optimization methods which drastically reduce the computation time to identify the cost-optimal static super-replication strategy. Furthermore, this computational advantage will allow us to also solve robustified versions of the hedging problem which is computationally not feasible for a Monte Carlo-based procedure.

Section 4.2 then makes use of the semi-infinite reformulation to explicitely derive and characterize the dual problem as the maximization over a specific class of probability measures. This insight allows to further compare the static super-replication approach with the well-known concept of dynamic super-replication and to interprete it from a financial point of view.

Without loss of generality, we restrict ourselves in the following to the interesting case of super-replicating an up-and-out call with strike K, barrier D, maturity $T > 0$ and payoff

$$C_{\text{uo}} = (S_T - K)^+ 1_{\{\tau > T\}}, \quad \tau := \inf\{t \in I : S_t = D\},$$

where τ denotes the stopping time of the first barrier hit and $\inf \emptyset := +\infty$.

4.1 Semi-infinite Equivalence

The super-replication problem (3.1) presented in Chapter 3 is a stochastic optimization problem defined for arbitrary financial market models and European options. In a first step, we restrict the set of models to a special class for which we want to prove the semi-infinite equivalence.

The results presented in Section 3.3 as well as the findings in [24, 27, 45, 69] show that European calls C^i with strike K_i and payoff $C^i(T_i, S_{T_i}) = (S_{T_i} - K_i)^+$ at maturity T_i are particularly suited hedge instruments to replicate an up-and-out call. Hence a model should be able to sufficiently approximate the prices of all standard calls in the hedge portfolio. As these calls differ in strikes and maturities it is natural to require a model used for static hedging to provide a good fit of the volatility surface. An example of such a surface of market data is presented in Figure 4.1 in the form of dots representing the implied volatilities of calls with several strikes and maturities. By definition the implied volatilities in the Black–Scholes model are a single constant which in turn leads to an extremely bad fit of the market data. This reveals that the

Black–Scholes model is not a good model for the derivation of static hedge portfolios for barrier options based on standard calls. In contrast to this Figure 4.1 shows that stochastic volatility models typically provide a good fit of the volatility surface.

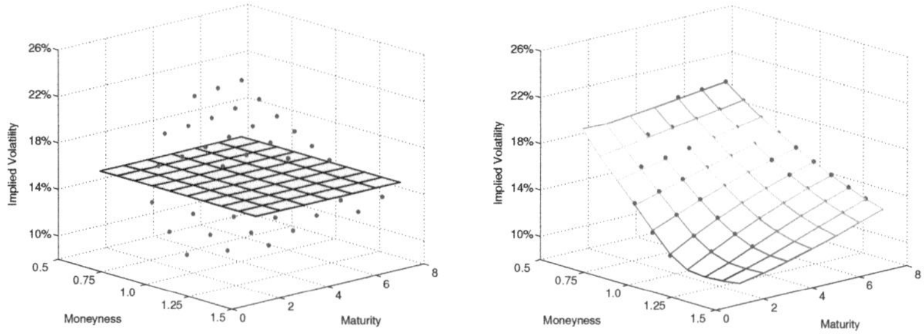

Figure 4.1: Market data (dots) and model fit for the Black–Scholes model (left) and a stochastic volatility model (right)

These findings motivate to choose stochastic volatility models for the derivation of cost-optimal static super-replicating strategies. The following definition specifies a broad class of these financial market models.

Assumption 4.1. Let $\mathcal{M}$ be a stochastic volatility model with time index set $I = [0, T]$, probability space $(\Omega, \mathcal{F}, P)$, bond B and stock price S, where the dynamics of the bond and stock price process under an equivalent martingale measure Q are given by the stochastic differential equations

$$\begin{aligned} dB_t &= rB_t dt, \; B_0 \in (0, \infty), \; r > 0 \\ dS_t &= (r - \delta)S_t dt + \sigma(Y_t)S_t dW_t^1, \; S_0 \in (0, \infty), \; \delta > 0 \\ dY_t &= \beta(Y_t)dt + \gamma(Y_t)dW_t^2, \; Y_0 \in \mathbb{R}. \end{aligned} \tag{4.1}$$

Here W^1, W^2 are two correlated Brownian motions with correlation coefficient ρ, $|\rho| < 1$, and the filtration $(\mathcal{F}_t)_{t\in[0,T]}$ is the augmented filtration generated by W^1, W^2. The process $(Y_t)_{t\in[0,T]}$ drives the volatility $\sigma(Y_t)$ of the stock price, where $\beta, \gamma : \mathbb{R} \to \mathbb{R}$ and $\sigma : \mathcal{Y} \to [0, \infty)$ are assumed to be measurable and $\mathcal{Y} \subset \mathbb{R}$ is a given interval with $Y_t \in \mathcal{Y} \; \forall \; t$ (a.s.). Furthermore, β and γ are assumed to satisfy the regularity conditions stated in Theorem 1.5 such that a solution $(Y_t)_{t\in[0,T]}$ of the second stochastic differential equation exists. Finally assume that σ is a continuous or bounded function.

The market model described above in particular contains Heston's stochastic volatility model and the Stein-Stein model as special cases, but also the simple Black–Scholes model and time-dependent volatility models can be obtained by an appropriate choice of the functions σ, β and γ.

As in Example 3.2 we now extend $\mathcal{M}$ by a set of tradable standard calls C^i, $i = 1, \ldots, n$, with strikes K_i and maturities $0 < T_i \leq T$. It is well known (see e.g. [41]) that under Assumption 4.1 the arbitrage-free value process $(C_t^i)_{t \in [0,T_i]}$ of these standard calls defined by

$$C_t^i := e^{-r(T_i - t)} E_Q((S_{T_i} - K_i)^+ | \mathcal{F}_t)$$

only depends on time t, the value of the stock price S_t and the process Y_t driving the volatility, that is $C_t^i = C^i(t, S_t, Y_t)$ with pricing function

$$C^i(t, s, y) := e^{-r(T_i - t)} E_Q(S_{T_i - t}^{s,y} - K_i)^+, \tag{4.2}$$

where $(S_\nu^{s,y})_{\nu \in [0, T_i - t]}$ is the solution of the stochastic differential equations (4.1) with initial conditions $S_0 = s \in (0, \infty)$ and $Y_0 = y \in \mathcal{Y}$. The resulting market model $\mathcal{M}$, extended by the calls C^i, $i = 1, \ldots, n$, then leads to the static super-replication problem

$$\begin{aligned}
&\min_{\alpha \in \mathbb{R}^{n+1}} \ \alpha_0 B_0 + \textstyle\sum_{i=1}^n \alpha_i C^i(0, S_0, Y_0) \\
\text{s.t.} \quad & \Pi_T(\alpha) = \alpha_0 B_T + \tfrac{B_T}{B_{\tau \wedge T}} \textstyle\sum_{i=1}^n \alpha_i C^i(\tau \wedge T, S_{\tau \wedge T}, Y_{\tau \wedge T}) \geq C_{\mathrm{uo}} \ \text{(a.s.)} \\
& C^i(t, S_t, Y_t) = \begin{cases} e^{-r(T_i - t)} E_Q(S_{T_i - t}^{S_t, Y_t} - K_i)^+, & 0 \leq t \leq T_i \\ C^i(T_i, S_{T_i}, Y_{T_i}) \frac{B_t}{B_{T_i}}, & T_i < t \leq T \end{cases} \\
& dS_t = (r - \delta) S_t dt + \sigma(Y_t) S_t dW_t^1 \\
& dY_t = \beta(Y_t) dt + \gamma(Y_t) dW_t^2 \\
& dB_t = r B_t dt,
\end{aligned} \tag{4.3}$$

where we made use of equation (3.7) to rewrite the portfolio value $\Pi_T(\alpha)$. Applying the general existence Theorem 3.3 we readily obtain that an optimal solution of problem (4.3) exists if a standard call with the same strike and maturity as the up-and-out call is included in the set of calls $\{C^i : i = 1, \ldots, n\}$.

In Section 3.2 we applied Algorithm 1 to numerically compute an optimal solution of problem (4.3). But as pointed out before, this Monte Carlo-based method has the significant drawback that the number of samples M has to be chosen quite large to obtain a good numerical approximation. As for each of the M samples the option prices $C^i(\tau \wedge T, S_{\tau \wedge T}, Y_{\tau \wedge T})$ and hence the conditional expectations have to be evaluated, the overall computation time for the solution of the problem can easily amount to several minutes or even hours – even if closed form solutions for the conditional expectations exist.

To reduce the computation time for the numerical solution of problem (4.3), we have to avoid the simulation of the sample paths of the stochastic differential equations for S_t and Y_t. But in general the value of the hedge portfolio Π_T will depend on the stopping time τ and the stock prices S_{T_i} at each maturity $T_i \leq \tau$ of the standard options. In this sense the value of a general static hedge portfolio for barrier options is path-dependent and hence the optimization problem has to be solved by a pathwise method like Algorithm 1. However, if we restrict the considered options to a reasonable choice, we can make use of the special structure of the problem to completely eliminate the stochastic nature.

Assumption 4.2. Assume that the calls C^i, $i = 1, \dots, n$, with strikes K_i and maturities $T_i \leq T$ satisfy $K_i \geq D$ in case $T_i < T$, where D is the barrier of the up-and-out call. Furthermore we assume that $K_i \neq K_j$ for $T_i = T_j, i \neq j$. The set of these calls shall be denoted by $\mathcal{C} = \{C^1, \dots, C^n\}$.

The last requirement $K_i \neq K_j$ for $T_i = T_j, i \neq j$ simply assures that all calls in the hedge portfolio either differ in strike or maturity. The more interesting assumption is the first one which states that options expiring before the up-and-out call have strikes greater or equal to the barrier. Calls with the same maturity as the barrier option may also have lower strikes $K_i \geq 0$. Intuitively it is clear that calls with strikes $K_i < D$ for $T_i < T$ are inefficient hedge instruments as they may result in large positive payoffs at time T_i while the up-and-out call expires worthless. By solving optimization problem (4.3) with the Monte Carlo-based Algorithm 1, we were also able to confirm numerically, that these inefficient calls are not part of a cost-optimal static hedge portfolio. In addition Assumption 4.2 is justified by the analytic static hedging results derived so far. For instance Derman, Ergener and Kani [27] as well as Carr, Ellis and Gupta [24] only consider hedge instruments (including binary calls) satisfying $K_i \geq D$ for $T_i < T$.

Restricting the calls in the portfolio to those described in Assumption 4.2, we can show that the value of a static hedge portfolio at time t is independent of the calls with maturities $T_i < t$ if the barrier has not been hit before time t.

Lemma 4.3. *Consider a financial market model $\mathcal{M}$ satisfying Assumption 4.1 and the static trading strategies, identified by $\alpha \in \mathbb{R}^{n+1}$, as stated in Definition 2.2 with corresponding portfolio value $\Pi_t(\alpha)$. Furthermore let $\omega \in \Omega$ and $t \in [0, T]$ be given such that $\tau(\omega) \geq t$. Then the value of a static hedge portfolio which consists of calls satisfying Assumption 4.2 is given by*

$$\Pi_t(\alpha)(\omega) = \alpha_0 B_t + \sum_{C^i \in \mathcal{C}, T_i \geq t} \alpha_i C^i(t, S_t(\omega), Y_t(\omega)).$$

Proof. By assumption we have $\tau(\omega) \geq t$ such that $S_v(\omega) < D \ \forall v \in [0, t)$. In particular this implies $S_{T_i}(\omega) < D$ for all calls $C^i \in \mathcal{C}$ with $T_i < t$. But due to Assumption 4.2 all these calls have strikes $K_i \geq D$ and hence must have expired worthless, because

$$C^i(T_i, S_{T_i}(\omega), Y_{T_i}(\omega)) = (S_{T_i}(\omega) - K_i)^+ \leq (D - K_i)^+ = 0.$$

This in turn implies $C^i(\nu, S_\nu(\omega), Y_\nu(\omega)) = 0\ \forall \nu \in [T_i, t]$ due to (3.4). Thus, by (3.7) we obtain

$$\begin{aligned}\Pi_t(\alpha)(\omega) &= \alpha_0 B_t + \frac{B_t}{B_t}\sum_{i=1}^{n} \alpha_i C^i(t, S_t(\omega), Y_t(\omega)) \\ &= \alpha_0 B_t + \sum_{C^i \in \mathcal{C}, T_i \geq t} \alpha_i C^i(t, S_t(\omega), Y_t(\omega)),\end{aligned}$$

which concludes the proof. □

Based on the representation of the hedge portfolio value in the previous lemma, we can now show that the stochastic constraint in problem (4.3) is equivalent to an infinite number of deterministic constraints.

Theorem 4.4. *Consider the problem (4.3) of finding the cost-optimal static super-replication strategy for a market model satisfying Assumption 4.1 and a set of calls as stated in Assumption 4.2. If the mappings $(t, y) \mapsto C^i(t, D, y)$ with C^i defined in (4.2) are continuous on $\operatorname{supp}(Q^{(\tau, Y_{\tau \wedge T})}) \cap ([0, T_i] \times \mathcal{Y})$, then the stochastic optimization problem (4.3) is equivalent to the deterministic linear semi-infinite optimization problem*

$$\begin{aligned}&\min_{\alpha \in \mathbb{R}^{n+1}} \alpha_0 B_0 + \sum_{i=1}^{n} \alpha_i C^i(0, S_0, Y_0) \\ \text{s.t.}\quad &\alpha_0 B_t + \sum_{C^i \in \mathcal{C}, T_i \geq t} \alpha_i C^i(t, D, y) \geq 0 \quad \forall\, (t, y) \in \Theta_1 \\ &\alpha_0 B_T + \sum_{C^i \in \mathcal{C}, T_i = T} \alpha_i (s - K_i)^+ \geq (s - K)^+ \quad \forall\, s \in \Theta_2,\end{aligned} \tag{4.4}$$

where Θ_1 and Θ_2 are given by $\Theta_1 = \operatorname{supp}(Q^{(\tau, Y_{\tau \wedge T})}) \cap ([0, T] \times \mathcal{Y})$ and $\Theta_2 = \{s \in [0, D] : (\infty, s) \in \operatorname{supp}(Q^{(\tau, S_T)})\}$.

Proof. It is sufficient to show that the stochastic constraint in (4.3) is equivalent to the infinite number of deterministic constraints in (4.4). We now distinguish the two cases of a barrier hit ($\tau \leq T$) and no barrier hit ($\tau = \infty$) until maturity T of the barrier option

$$\Pi_T(\alpha) \geq C_{\text{uo}} \text{ (a.s.)} \quad \Longleftrightarrow \quad \begin{cases} \Pi_{\tau \wedge T}(\alpha) 1_{\{\tau \leq T\}} \geq 0 \text{ (a.s.)} \\ \Pi_T(\alpha) 1_{\{\tau = \infty\}} \geq C_{\text{uo}} 1_{\{\tau = \infty\}} \text{ (a.s.)}, \end{cases}$$

where we used for the case $\tau \leq T$ that due to (3.7) $\Pi_T(\alpha) = \Pi_{\tau \wedge T}(\alpha) \cdot B_T / B_{\tau \wedge T}$. Applying Lemma 4.3 to both cases then shows that $\Pi_T(\alpha) \geq C_{\text{uo}}$ (a.s.) is equivalent to the set of inequalities

$$\left(\alpha_0 B_{\tau\wedge T} + \sum_{C^i \in \mathcal{C}, T_i \geq \tau\wedge T} \alpha_i C^i(\tau \wedge T, S_{\tau\wedge T}, Y_{\tau\wedge T})\right) 1_{\{\tau \leq T\}} \geq 0 \text{ (a.s.)}$$

$$\left(\alpha_0 B_T + \sum_{C^i \in \mathcal{C}, T_i = T} \alpha_i C^i(T, S_T, Y_T)\right) 1_{\{\tau=\infty\}} \geq (S_T - K)^+ 1_{\{\tau=\infty\}} \text{ (a.s.)}.$$

Note that for $\tau \leq T$ we have $S_{\tau\wedge T} = S_\tau = D$ and hence $C^i(\tau \wedge T, S_{\tau\wedge T}, Y_{\tau\wedge T}) = C^i(\tau \wedge T, D, Y_{\tau\wedge T})$. Furthermore, for $T_i = T$ the value of call C^i at time T is given by $C^i(T, S_T, Y_T) = (S_T - K_i)^+$. Hence $\Pi_T(\alpha) \geq C_{\mathrm{uo}}$ (a.s.) is equivalent to

$$\left(\alpha_0 B_{t\wedge T} + \sum_{C^i \in \mathcal{C}, T_i \geq t\wedge T} \alpha_i C^i(t \wedge T, D, y)\right) 1_{\{t \leq T\}} \geq 0 \, \left(Q^{(\tau, Y_{\tau\wedge T})} - \text{a.s.}\right)$$

$$\left(\alpha_0 B_T + \sum_{C^i \in \mathcal{C}, T_i = T} \alpha_i (s - K_i)^+\right) 1_{\{t=\infty\}} \geq (s - K)^+ 1_{\{t=\infty\}} \, \left(Q^{(\tau, S_T)} - \text{a.s.}\right).$$

We now focus on the first case $t \leq T$. The proof of the second case $t = \infty$ follows in analogy. It remains to show that

$$\left(\alpha_0 B_{t\wedge T} + \sum_{C^i \in \mathcal{C}, T_i \geq t\wedge T} \alpha_i C^i(t \wedge T, D, y)\right) 1_{\{t \leq T\}} \geq 0 \, \left(Q^{(\tau, Y_{\tau\wedge T})} - \text{a.s.}\right) \iff$$

$$\alpha_0 B_t + \sum_{C^i \in \mathcal{C}, T_i \geq t} \alpha_i C^i(t, D, y) \geq 0 \quad \forall\, (t, y) \in \operatorname{supp}(Q^{(\tau, Y_{\tau\wedge T})}) \cap ([0, T] \times \mathcal{Y}).$$

The direction "$\Leftarrow$" immediately follows by the definition of the support of a measure. For the proof of the other direction "$\Rightarrow$" we first observe that

$$\left(\alpha_0 B_{t\wedge T} + \sum_{C^i \in \mathcal{C}, T_i \geq t\wedge T} \alpha_i C^i(t \wedge T, D, y)\right) 1_{\{t \leq T\}} \geq 0 \, \left(Q^{(\tau, Y_{\tau\wedge T})} - \text{a.s.}\right)$$

$$\iff \quad \alpha_0 B_t + \textstyle\sum_{C^i \in \mathcal{C}, T_i \geq t} \alpha_i C^i(t, D, y) \geq 0 \quad \left(Q^{(\tau, Y_{\tau\wedge T})}|_{[0,T]\times\bar{\mathcal{Y}}} - \text{a.s.}\right).$$

This implies that there exists a set N with $Q^{(\tau, Y_{\tau\wedge T})}|_{[0,T]\times\bar{\mathcal{Y}}}(N) = 0$ such that

$$g(t, y) := \alpha_0 B_t + \sum_{C^i \in \mathcal{C}, T_i \geq t} \alpha_i C^i(t, D, y) \geq 0 \quad \forall\, (t, y) \in N^c.$$

If we define $M := N^c \cap \operatorname{supp}(Q^{(\tau, Y_{\tau\wedge T})}|_{[0,T]\times\bar{\mathcal{Y}}})$ the latter inequality also holds on M which is a set of measure one. If g is continuous on $\operatorname{supp}(Q^{(\tau, Y_{\tau\wedge T})}|_{[0,T]\times\bar{\mathcal{Y}}})$, then $g(t, y) \geq 0$ on the closure of M denoted by $\bar{M}$. As $\bar{M}$ is a closed set with measure one, $g(t, y) \geq 0$ must also hold on the smallest closed set with measure one which is by definition $\operatorname{supp}(Q^{(\tau, Y_{\tau \wedge T})}|_{[0,T]\times\bar{\mathcal{Y}}})$. Finally it is easy to show that

$$\operatorname{supp}(Q^{(\tau, Y_{\tau \wedge T})}|_{[0,T]\times\bar{\mathcal{Y}}}) = \operatorname{supp}(Q^{(\tau, Y_{\tau \wedge T})}) \cap \left([0, T] \times \bar{\mathcal{Y}}\right),$$

which proves the theorem.

Regarding the continuity of g note that by assumption the maps $(t, y) \mapsto C^i(t, D, y)$ are continuous on $\operatorname{supp}\big(Q^{(\tau, Y_{\tau \wedge T})}\big) \cap \left([0, T_i] \times \bar{\mathcal{Y}}\right)$. This immediately implies the continuity of g in points (t, y) with $t \neq T_i < T$. However, for $t = T_i < T$ the index set of the summation differs to the left and right of T_i which might destroy the continuity in these points. But note that for a sequence $(t_k, y_k)_k$ converging to (T_i, y) with $t_k < T_i \ \forall k \in \mathbb{N}$ it follows that

$$C^i(t_k, D, y_k) \rightarrow_{k\rightarrow\infty} C^i(T_i, D, y) = (D - K_i)^+ = 0$$

due to $K_i \geq D$. Thus g is also continuous in points (T_i, y) and hence on the entire set $\operatorname{supp}(Q^{(\tau, Y_{\tau \wedge T})}) \cap \left([0, T] \times \bar{\mathcal{Y}}\right) = \operatorname{supp}(Q^{(\tau, Y_{\tau \wedge T})}|_{[0,T]\times\bar{\mathcal{Y}}})$. □

Although Theorem 4.4 looks complicated on first sight, the result allows an easy and intuitive interpretation. The original stochastic constraint $\Pi_T(\alpha) \geq C_{\mathrm{uo}}$ (a.s.) of problem (4.3) requires the hedge portfolio to super-replicate the up-and-out call in all possible states of the postulated financial market model. Note that this model is a stochastic volatility model, hence the states of the model are described by the two stochastic differential equations (4.1) for S_t and Y_t. These states are now split into those cases in which the barrier is hit until maturity and the cases in which the barrier is not hit at all. Of course the super-replication property has to hold for both cases.

On the one hand, in case the barrier is hit ($\tau \leq T$), the hedge portfolio only consists of calls with maturities greater or equal to the time of the barrier hit as all calls with shorter maturities T_i have expired worthless (Lemma 4.3). The remaining calls have to super-replicate the value of the barrier option $C_{\mathrm{uo}} 1_{\{\tau \leq T\}} = 0$ in all possible states of the market model for which the barrier is hit. Hence the super-replication property has to be guaranteed for all time-volatility combinations $(\tau, Y_{\tau \wedge T})$ with $\tau \leq T$. These combinations are given by the set $\Theta_1 = \operatorname{supp}\big(Q^{(\tau, Y_{\tau \wedge T})}\big) \cap \left([0, T] \times \bar{\mathcal{Y}}\right)$.

On the other hand, in case the barrier is not hit at all ($\tau = \infty$), the super-replication property requires the payoff of all remaining calls in the portfolio at time T to be greater or equal to the value of the up-and-out call $C_{\mathrm{uo}} 1_{\{\tau=\infty\}} = (S_T - K)^+$. These calls have maturities $T_i = T$ and hence their payoff at time T does not depend upon the volatility at time T, but only on the stock price S_T at terminal time. Accordingly, the calls have to super-replicate the barrier option for all possible stock prices S_T which can be attained if the barrier is not hit. The set of these states is given by $\Theta_2 = \{s \in [0, D] : (\infty, s) \in \operatorname{supp}(Q^{(\tau, S_T)})\}$.

Due to Theorem 4.4 a Monte Carlo type simulation of the stochastic differential equations for S_t and Y_t is not necessary for the solution of optimization problem (4.3).

The states for which the super-replication property has to hold are known in advance and given by Θ_1 and Θ_2. In general these sets can describe very complicated subsets of $\mathbb{R}^2$ and $\mathbb{R}$, respectively. However, for most stochastic volatility models Θ_1 and Θ_2 have a rather simple interval-type structure as the following example shows.

Example 4.5. We now state the sets Θ_1 and Θ_2 for four famous models which are special cases of the general model defined in Assumption 4.1, namely the Black–Scholes model, time-dependent volatility models, Heston's stochastic volatility model and the Stein-Stein model. We omit detailed proofs at this point and rather focus on the explanation of the underlying intuitive idea.

i) The Black–Scholes model is given by the model equations

$$\begin{aligned} dS_t &= (r-\delta)S_t dt + \sqrt{Y_t} S_t dW_t^1, \ S_0 \in (0,\infty) \\ dY_t &= 0dt + 0dW_t^2, \ Y_0 \in (0,\infty), \end{aligned}$$

which is a special case of the general market model (4.1) if we choose $\beta \equiv 0$ and $\gamma \equiv 0$. Hence $Y_t \equiv Y_0 =$ const such that the volatility $\sigma(Y_t) = \sqrt{Y_0} =: \sigma_0$ of the stock price is constant. Clearly the stock price can hit the barrier at any time $t \in (0,T]$ such that the support of the stopping time $\tau \wedge T$ is $[0,T]$. As the process $Y_t = Y_0$ driving the volatility is constant it also equals Y_0 in case the barrier is hit. Thus $Y_{\tau\wedge T} = Y_0$ and hence $\Theta_1 = \operatorname{supp}\big(Q^{(\tau, Y_{\tau\wedge T})}\big) \cap \big([0,T] \times \bar{\mathcal{Y}}\big) = [0,T] \times \{Y_0\}$.

ii) In time-dependent volatility models the assumption of a constant volatility is relaxed to a deterministic volatility as described by the model equations

$$\begin{aligned} dS_t &= (r-\delta)S_t dt + \sigma(Y_t) S_t dW_t^1, \ S_0 \in (0,\infty) \\ dY_t &= \beta(Y_t)dt + 0dW_t^2, \ Y_0 \in (0,\infty). \end{aligned}$$

Now the process $Y_t = Y(t)$ is the solution of a deterministic differential equation such that $\sigma(Y_t) = \sigma(t)$ is also a deterministic function. Still the barrier can be hit any time $t \in (0,T]$ (if $\sigma(t) > 0 \ \forall t$) with the corresponding source of volatility $Y(t)$. Thus we have $\Theta_1 = \operatorname{supp}\big(Q^{(\tau, Y_{\tau\wedge T})}\big) \cap \big([0,T] \times \bar{\mathcal{Y}}\big) = \{(t, Y(t)) : t \in [0,T]\}$.

iii) To account for the stochastic nature of the volatility of the stock price, Heston [50] incorporated an additional source of uncertainty in the process driving the volatility function

$$\begin{aligned} dS_t &= (r-\delta)S_t dt + \sqrt{Y_t} S_t dW_t^1, \ S_0 \in (0,\infty) \\ dY_t &= \kappa(\theta - Y_t)dt + \xi\sqrt{Y_t} dW_t^2, \ Y_0 \in (0,\infty), \end{aligned}$$

where κ, θ and ξ are positive constants satisfying $\kappa\theta - \xi^2/2 \geq 0$. As Cox, Ingersoll and Ross [26] note, the distribution of Y_t given Y_s for some $s < t$ is, up to a scale factor, a noncentral chi-squared distribution and hence has a support of $[0,\infty)$. Furthermore Broadie and Kaya [18] prove that the conditional distribution of $\log(S_t)$ given $(Y_s)_{0\leq s\leq T}$ is normal. Hence the barrier can be hit any time $t \in (0,T]$ and at that time the process Y_t can attain any value $Y_t \in [0, \infty)$. This implies that for Heston's model $\Theta_1 = \text{supp}\big(Q^{(\tau, Y_{\tau \wedge T})}\big) \cap \big([0,T] \times \bar{\mathcal{Y}}\big) = [0,T] \times [0, \infty)$.

iv) The Stein-Stein model (see e.g. [41]) is another well-known stochastic volatility model driven by the model equations

$$\begin{aligned} dS_t &= (r - \delta)S_t dt + |Y_t| S_t dW_t^1, \ S_0 \in (0, \infty) \\ dY_t &= \kappa(\theta - Y_t)dt + \xi dW_t^2, \ Y_0 \in (0, \infty) \end{aligned}$$

with constants $\kappa, \theta, \xi > 0$. Here Y_t is a Gaussian process such that Y_t can attain any value in $\mathbb{R}$ in case the barrier is hit. Accordingly, we obtain in this model $\Theta_1 = \text{supp}\big(Q^{(\tau, Y_{\tau \wedge T})}\big) \cap \big([0,T] \times \bar{\mathcal{Y}}\big) = [0,T] \times (-\infty, \infty)$.

As a summary, Figure 4.2 illustrates the time-volatility combinations $(t, y) \in \Theta_1$ for which the hedge portfolio has to super-replicate the up-and-out call. Regarding the set $\Theta_2 = \big\{s \in [0, D] : (\infty, s) \in \text{supp}\big(Q^{(\tau, S_T)}\big)\big\}$, that is the set of all possible stock prices S_T in case the barrier is not hit at all, we obtain $\Theta_2 = [0, D]$ for all four models.

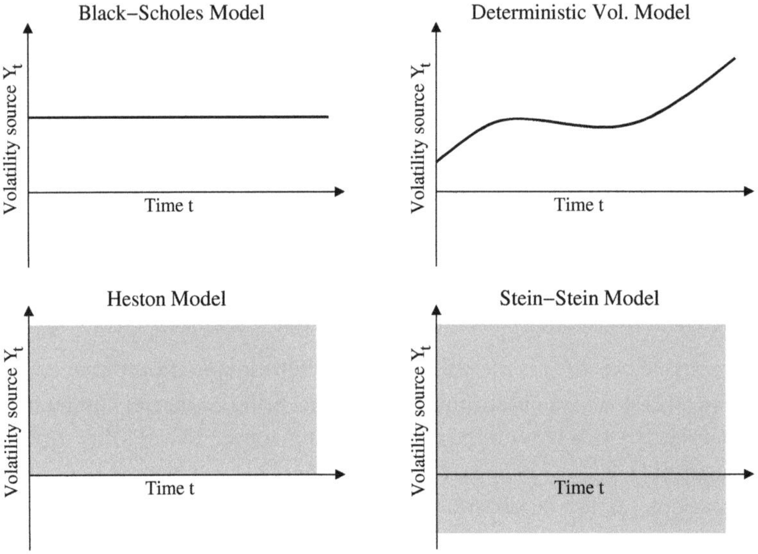

Figure 4.2: Sketch of the set $\Theta_1 := \text{supp}\big(Q^{(\tau, Y_{\tau \wedge T})}\big) \cap \big([0,T] \times \bar{\mathcal{Y}}\big)$ for typical financial market models

Based on the derivations above it is clear that a cost-optimal super-replication strategy will crucially depend on the support of the measures $Q^{(\tau, Y_{\tau \wedge T})}$ and $Q^{(\tau, S_T)}$. Other static hedging approaches like the minimization of the L^2 hedge error (see Subsection 2.3.3) by definition focus on expected value functionals which are computationally easier to handle but neglect low probability events. This leads to possibly huge hedge errors for the unlikely but dangerous cases of barrier hits close to maturity of the barrier option. In contrast to this the support-based approach pursued in this work also guarantees protection for unlikely but critical events.

Furthermore, as we will see in Chapter 5, the deterministic semi-infinite reformulation of the stochastic optimization problem (4.3) can be exploited by specialized algorithms. But before we derive appropriate methods for the solution of problem (4.4), we first utilize its structure to analyze the dual optimization problem.

4.2 The Dual Problem

The study of the dual of an optimization problem usually leads to additional insights regarding the structure of the problem as well as its economic interpretation. In the context of super-replication it is well known (see e.g. El-Karoui and Quenez [30]), that the minimization of the portfolio cost over the space of all dynamic trading strategies super-replicating the target option payoff can be rephrased as the dual problem of maximizing the discounted expected payoff of the target option over the set of all equivalent martingale measures.

This raises the question what the dual problem of the static super-replication problem (4.3) looks like. In comparison to the dynamic super-replication problem, the feasible set of static trading strategies is much smaller. In fact the static strategies only form a finite dimensional subspace of the set of all dynamic trading strategies. Correspondingly, the feasible set of the dual problem for the static case will be much bigger than for the dual of the dynamic problem. This means that in the dual static problem the class of admissible probability measures will consist of a lot more elements than the equivalent martingale measures.

While these basic insights are intuitively clear, the derivation of the dual problem and the precise characterization of the feasible measures is still an open question. In this section we will present a duality result of Maruhn [67], which shows that the deterministic semi-infinite representation (4.4) of problem (4.3) can be used to set up the dual problem and answer the questions above. As it turns out, the feasible set of the dual problem is given by the measures consistent with the prices of the calls included in the hedge portfolio. Furthermore we prove that there exists an optimal measure solving the dual problem which is the weighted sum of mostly $n + 1$ Dirac measures. This in turn allows to further analyze and interprete the static hedging problem from a financial point of view.

To facilitate the derivation of the dual problem we focus on the case of super-replicating an up-and-out call in the Black–Scholes model. In this model the super-replication problem (4.3) can be rephrased (see Theorem 4.4 and Example 4.5) as the linear semi-infinite optimization problem

$$\min_{\alpha \in \mathbb{R}^{n+1}} \alpha_0 B_0 + \sum_{i=1}^{n} \alpha_i C^i(0, S_0, \sigma_0^2)$$

$$\text{s.t.} \qquad \alpha_0 B_t + \sum_{C^i \in \mathcal{C}, T_i \geq t} \alpha_i C^i(t, D, \sigma_0^2) \geq 0 \quad \forall\, t \in [0, T] \tag{4.5}$$
$$\alpha_0 B_T + \sum_{C^i \in \mathcal{C}, T_i = T} \alpha_i (s - K_i)^+ \geq (s - K)^+ \quad \forall\, s \in [0, D],$$

where the call prices $C^i(t, s, \sigma_0^2)$, given the constant volatility σ_0, can be computed via the Black–Scholes formula (see (1.10))

$$\begin{aligned} C^i(t, s, \sigma_0^2) = {} & s e^{-\delta(T_i - t)} \Phi\left(\frac{\log(\frac{s}{K_i}) + (r - \delta + \frac{\sigma_0^2}{2})(T_i - t)}{\sigma_0 \sqrt{T_i - t}} \right) \\ & - K_i e^{-r(T_i - t)} \Phi\left(\frac{\log(\frac{s}{K_i}) + (r - \delta - \frac{\sigma_0^2}{2})(T_i - t)}{\sigma_0 \sqrt{T_i - t}} \right) \end{aligned} \tag{4.6}$$

and $\Phi(\cdot)$ denotes the cumulative normal distribution. If we define the mappings g^i : $[0, T] \times \{D\} \cup \{T\} \times [0, D] \to \mathbb{R}$,

$$\text{For } T_i < T: \quad g^i(t, s) := \begin{cases} C^i(t, D, \sigma_0^2), & 0 \leq t \leq T_i \\ 0, & T_i < t < T \\ 0, & t = T,\ 0 \leq s \leq D \end{cases} \tag{4.7}$$

$$\text{For } T_i = T: \quad g^i(t, s) := \begin{cases} C^i(t, D, \sigma_0^2), & 0 \leq t < T \\ (s - K_i)^+, & t = T,\ 0 \leq s \leq D, \end{cases} \tag{4.8}$$

as well as $h : [0, T] \times \{D\} \cup \{T\} \times [0, D] \to \mathbb{R}$ by

$$h(t, s) := \begin{cases} 0, & 0 \leq t < T \\ (s - K)^+, & t = T,\ 0 \leq s \leq D, \end{cases}$$

then problem (4.5) is equivalent to

$$\begin{aligned} & \min_{\alpha \in \mathbb{R}^{n+1}} \ \alpha_0 B_0 + \sum_{i=1}^{n} \alpha_i C^i(0, S_0, \sigma_0^2) \\ & \text{s.t.} \quad \alpha_0 B_t + \sum_{i=1}^{n} \alpha_i g^i(t, s) \geq h(t, s) \quad \forall\, (t, s) \in L, \end{aligned} \tag{4.9}$$

where $L := [0, T] \times \{D\} \cup \{T\} \times [0, D]$ is a set with L-shape (see Figure 4.3 for a graphical illustration). Hence the constraints can be interpreted as a one-sided approximation of the function h on L by a linear combination of the functions B and g^i. This motivates to further analyze the properties of those functions.

Lemma 4.6. *Under Assumption 4.2 the functions g^i defined in (4.7) and (4.8) are continuous on $L := [0, T] \times \{D\} \cup \{T\} \times [0, D]$.*

Proof. For the case $T_i = T$ the continuity immediately follows due to the continuity of the Black–Scholes formula (4.6). To verify that g^i is also continuous for the case $T_i < T$, it is sufficient to show that the left hand limit $\lim_{t \to T_i -} C^i(t, D, \sigma_0^2)$ equals zero. Note that

$$\lim_{t \to T_i -} \Phi\left(\frac{\log(D/K_i) + (r - \delta \pm \sigma_0^2/2)(T_i - t)}{\sigma_0 \sqrt{T_i - t}} \right) =$$

$$= \lim_{t \to T_i -} \Phi\left(\frac{\log(D/K_i)}{\sigma_0 \sqrt{T_i - t}} \right) = \begin{cases} \Phi(0) = \frac{1}{2}, & K_i = D \\ \Phi(-\infty) = 0, & K_i > D. \end{cases}$$

This in turn implies

$$\lim_{t \to T_i -} C^i(t, D, \sigma_0^2) = \begin{cases} D \cdot \frac{1}{2} - K_i \cdot \frac{1}{2} = 0, & K_i = D \\ D \cdot 0 - K_i \cdot 0 = 0, & K_i > D, \end{cases}$$

which concludes the proof. □

In contrast to the functions g^i the barrier option payoff h along L is discontinuous in (T, D). Combined with the fact that the mapping $(t, s) \mapsto B_t$ is trivially continuous on L, we can deduce from Lemma 4.6 that the constraints of problem (4.9) can in fact be interpreted as approximating the discontinuous barrier option on L by a linear combination of continuous bond and call values. This interpretation is also graphically visible in Figure 4.3 which illustrates the function h as well as its one-sided approximation by the static super-replication portfolio computed in Subsection 3.3.1. Hence the derivations above now formally prove that the static super-replication approach automatically regularizes the hedging problem for barrier options by closing the continuity gap with the best achievable continuous approximation.

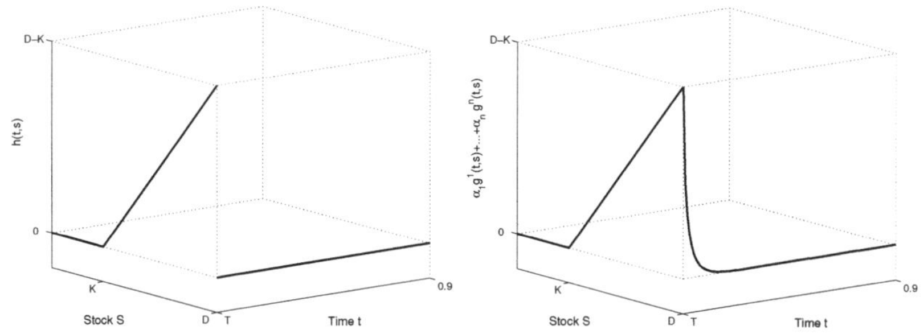

Figure 4.3: Graphical illustration of the function h (left) on the set L and its one-sided continuous approximation (right) by the optimal super-replication portfolio listed in Table 3.1

In the remainder of this section the derived continuity of the functions g^i will further allow us to obtain a nice representation of the dual of the static hedging problem. As a first step towards this goal we prove that problem (4.5) is in fact a one-sided L^1-approximation problem.

Theorem 4.7. *The static super-replication problem (4.5) is equivalent to the one-sided L^1-approximation problem*

$$\min_{\alpha\in\mathbb{R}^{n+1}} \alpha_0 \int_L B_t e^{-rt}\, dQ^{(\tau\wedge T, S_{\tau\wedge T})} + \sum_{i=1}^{n} \alpha_i \int_L g^i(t,s) e^{-rt}\, dQ^{(\tau\wedge T, S_{\tau\wedge T})}$$

$$\text{s.t.} \quad \alpha_0 B_t e^{-rt} + \sum_{i=1}^{n} \alpha_i g^i(t,s) e^{-rt} \geq h(t,s) e^{-rt} \tag{4.10}$$

$$\forall\, (t,s) \in L := [0,T]\times\{D\} \cup \{T\}\times[0,D].$$

Proof. Based on the derivation of (4.9) it is sufficient to prove that

$$B_0 = \int_L B_t e^{-rt}\, dQ^{(\tau\wedge T, S_{\tau\wedge T})},\ C^i(0, S_0, \sigma_0^2) = \int_L g^i(t,s) e^{-rt}\, dQ^{(\tau\wedge T, S_{\tau\wedge T})}.$$

The first equality immediately follows if we observe that $B_t = B_0 e^{rt}$. In order to prove the second equality we distinguish the cases $T_i < T$ and $T_i = T$.

For the case $T_i < T$ Assumption 4.2 implies $K_i \geq D$ and hence we obtain for $t > T_i$ that $g^i(t,s) = 0 = (D - K_i)^+ = C^i(T_i, D, \sigma_0^2) = C^i(t\wedge T_i, D, \sigma_0^2)$. Thus we can derive

$$\begin{aligned}
\int_L g^i(t,s) e^{-rt}\, dQ^{(\tau\wedge T, S_{\tau\wedge T})} &= \int_L C^i(t\wedge T_i, D, \sigma_0^2) e^{-r(t\wedge T_i)}\, dQ^{(\tau\wedge T, S_{\tau\wedge T})} \\
&= \int_\Omega C^i(\tau\wedge T_i, D, \sigma_0^2) e^{-r(\tau\wedge T_i)}\, dQ \\
&= e^{-rT_i} \int_\Omega C^i(\tau\wedge T_i, D, \sigma_0^2) \frac{B_{T_i}}{B_{\tau\wedge T_i}}\, dQ.
\end{aligned}$$

The same result can be obtained for the case $T_i = T$ by observing that

$$\begin{aligned}
&\int_L g^i(t,s) e^{-rt}\, dQ^{(\tau\wedge T, S_{\tau\wedge T})} \\
&= \int_{[0,T)\times\{D\}} C^i(t, D, \sigma_0^2) e^{-rt}\, dQ^{(\tau\wedge T, S_{\tau\wedge T})} \\
&\qquad + \int_{\{T\}\times[0,D]} (s - K_i)^+ e^{-rt}\, dQ^{(\tau\wedge T, S_{\tau\wedge T})} \\
&= \int_\Omega C^i(\tau\wedge T, D, \sigma_0^2) e^{-r(\tau\wedge T)} 1_{\{\tau<T\}}\, dQ + \int_\Omega (S_T - K_i)^+ e^{-rT} 1_{\{\tau\geq T\}}\, dQ \\
&= e^{-rT} \int_\Omega C^i(\tau\wedge T, D, \sigma_0^2) \frac{B_T}{B_{\tau\wedge T}} 1_{\{\tau<T\}}\, dQ \\
&\qquad + e^{-rT} \int_\Omega C^i(\tau\wedge T, D, \sigma_0^2) \frac{B_T}{B_{\tau\wedge T}} 1_{\{\tau\geq T\}}\, dQ \\
&= e^{-rT} \int_\Omega C^i(\tau\wedge T_i, D, \sigma_0^2) \frac{B_{T_i}}{B_{\tau\wedge T_i}}\, dQ.
\end{aligned}$$

But the latter expression is just the discounted expected value of the financial product with payoff $C^i(\tau(\omega) \wedge T_i, D, \sigma_0^2)B_{T_i}/B_{\tau(\omega)\wedge T_i}$ at time T_i which coincides with the payoff of the strategy of buying one unit of call C^i and shifting its value to the bond position as soon as the barrier is hit or the option expires. As the Black–Scholes model, extended by the value processes of the standard calls C^i, is arbitrage-free, the value of the latter strategy must be equal to the value of the call at time $t = 0$, i.e. $C^i(0, S_0, \sigma_0^2)$. □

Based on Theorem 4.7 the idea arises to derive the dual of the static super-replication problem by interpreting it as a one-sided L^1-approximation problem. If both sides of the inequality in (4.10) would be continuous on the compact domain L and if $Q^{(\tau\wedge T, S_{\tau\wedge T})}$ would be a strictly positive measure on L, that is

$$\int_L f(t,s)dQ^{(\tau\wedge T, S_{\tau\wedge T})} > 0 \;\; \forall \text{ continuous functions } f : L \to \mathbb{R},\; f \geq 0,\; f \neq 0,$$

then we could readily deduce the dual of problem (4.10) by applying appropriate results from the theory of semi-infinite optimization (see Theorem 1.15). But while the requirement on the measure $Q^{(\tau\wedge T, S_{\tau\wedge T})}$ is fulfilled due to $\text{supp}(Q^{(\tau\wedge T, S_{\tau\wedge T})}) = L$, the discontinuity of the barrier option payoff h poses a problem. However, the next theorem shows that it is still possible to derive the dual optimization problem including a nice characterization of the set of feasible measures.

Theorem 4.8. *The dual of the static hedging problem (4.5) is given by*

$$\begin{aligned} &\max_{\mu\in\mathbb{P}(L)} \int_L e^{-rt} h(t,s)d\mu(t,s) \\ \text{s.t.} \quad &\int_L e^{-rt} g^i(t,s)d\mu(t,s) = C^i(0, S_0, \sigma_0^2) \quad \forall\, i = 1,\dots,n, \end{aligned} \tag{D}$$

where $\mathbb{P}(L)$ denotes the set of all probability measures μ on $L \subset \mathbb{R}^2$ and the corresponding Borel sigma algebra. In particular problem (D) has a solution and the optimal values of (4.5) and (D) coincide.

Moreover there exists an optimal solution of the dual problem (D) in the form of a discrete measure $\mu^* = \sum_{l=1}^d w_l \delta_{\{t_l,s_l\}}$, $d \in \mathbb{N}$, $d \leq n+1$, $0 \leq w_l \leq 1$, $\sum_{l=1}^d w_l = 1$, which is the weighted sum of d Dirac measures $\delta_{\{t_l,s_l\}}$ at the points $(t_l, s_l) \in L$ in which the constraints of (4.10) are active for a given optimal super-replication portfolio α^* solving (4.5), (4.10).

Proof. First we observe that the payoff of the up-and-out call C_{uo} is bounded from above by $(D-K)^+$ and hence it is super-replicated by choosing a sufficiently large bond position α_0. Thus Theorem 3.3 implies that a solution of the super-replication problem (4.5) exists. In the following let $v(P) > -\infty$ denote the optimal function value and let $\alpha^* = (\alpha_0^*, \alpha_1^*, \ldots, \alpha_n^*)^{\mathrm{T}}$ be an optimal solution. By Theorem 4.7, α^* also solves the one-sided L^1-approximation problem

$$\begin{aligned} &\min_{\alpha \in \mathbb{R}^{n+1}} \ \alpha_0 B_0 + \sum_{i=1}^{n} \alpha_i C^i(0, S_0, \sigma_0^2) \\ &\text{s.t.} \quad \alpha_0 B_t e^{-rt} + \sum_{i=1}^{n} \alpha_i g^i(t,s) e^{-rt} \geq h(t,s) e^{-rt} \quad \forall\,(t,s) \in L, \end{aligned} \tag{P}$$

implying in particular $\alpha_0^* B_t + \sum_{i=1}^{n} \alpha_i^* g^i(t,s) \geq h(t,s)\ \forall (t,s) \in L$. Note that $h(T,D) = (D-K)^+ > 0$ such that

$$\begin{aligned} \alpha_0^* B_T + \sum_{i=1}^{n} \alpha_i^* g^i(T,D) &> 0 \\ \alpha_0^* B_t + \sum_{i=1}^{n} \alpha_i^* g^i(t,D) &\geq 0 \quad \forall t \in [0,T]. \end{aligned} \tag{4.11}$$

As proven in Lemma 4.6, the left hand sides of the inequalities (4.11) are continuous functions on L. Hence there exists a time interval $(T-\epsilon, T]$ such that $\alpha_0^* B_t + \sum_{i=1}^{n} \alpha_i^* g^i(t,D)$ is strictly positive for all $t \in (T-\epsilon, T]$. The idea is now to construct a continuous modification of h on this time interval without changing the optimal value $v(P)$ of the primal problem (P). Defining the function

$$u_\epsilon(t) := \begin{cases} 0, & 0 \leq t \leq T-\epsilon \\ \frac{1}{\epsilon}(t-(T-\epsilon)) \frac{(D-K)^+}{\alpha_0^* B_T + \sum_{i=1}^{n} \alpha_i^* g^i(T,D)}, & T-\epsilon < t \leq T \end{cases}$$

the desired continuous modification of h is given by

$$h_\epsilon(t,s) := \begin{cases} 0, & 0 \leq t \leq T-\epsilon \\ u_\epsilon(t)\{\alpha_0^* B_t + \sum_{i=1}^{n} \alpha_i^* g^i(t,D)\}, & T-\epsilon < t < T \\ (s-K)^+, & t = T,\ s \in [0,D]. \end{cases}$$

By definition $h_\epsilon(t,s) \geq h(t,s)\ \forall\,(t,s) \in L$ and due to $0 \leq u_\epsilon(t) \leq 1\ \forall\, t$ as well as $0 < u_\epsilon(t) < 1, t \in (T-\epsilon, T)$, it follows by (4.11) and the choice of ϵ that

$$\begin{aligned} \alpha_0^* B_t + \sum_{i=1}^{n} \alpha_i^* g^i(t,s) &\geq h_\epsilon(t,s) \quad \forall (t,s) \in L \\ \alpha_0^* B_t + \sum_{i=1}^{n} \alpha_i^* g^i(t,s) &> h_\epsilon(t,s) \quad \forall (t,s) \in (T-\epsilon, T) \times \{D\}. \end{aligned} \tag{4.12}$$

This shows that α^* is also feasible for the one-sided L^1-approximation problem

$$\begin{aligned} &\min_{\alpha\in\mathbb{R}^{n+1}} \alpha_0 B_0 + \sum_{i=1}^{n} \alpha_i C^i(0, S_0, \sigma_0^2) \\ \text{s.t.}\quad &\alpha_0 B_t e^{-rt} + \sum_{i=1}^{n} \alpha_i g^i(t,s) e^{-rt} \geq h_\epsilon(t,s) e^{-rt} \quad \forall\,(t,s) \in L, \end{aligned} \tag{P_ϵ}$$

which has less feasible points than (P) due to $h_\epsilon(t,s) \geq h(t,s)$. Hence α^* is also an optimal solution of (P_ϵ) implying that the optimal values $\nu(P)$ of (P) and $\nu(P_\epsilon)$ of (P_ϵ) coincide. Further problem (P_ϵ) satisfies the slater condition, because we can choose a sufficiently large bond position $\alpha_0 = (M+1)/B_0$, $M := \max_{(t,s)\in L} h_\epsilon(t,s)$, and $\alpha_1 = \ldots = \alpha_n = 0$ such that

$$\alpha_0 B_t + \sum_{i=1}^{n} \alpha_i g^i(t,s) = (M+1)e^{rt} > M \geq h_\epsilon(t,s) \quad \forall\,(t,s) \in L.$$

But for linear semi-infinite optimization problems with these properties it is well known (see for example Hettich [52], Theorem 3.2.13) that the strong duality theorem holds. Furthermore, as (P_ϵ) is a one-sided L^1-approximation problem with continuous functions, the dual problem is known (see Theorem 1.15) to be given by

$$\begin{aligned} &\max_{\mu} \int_L e^{-rt} h_\epsilon(t,s)\,d\mu(t,s) \\ \text{s.t.}\quad &\int_L e^{-rt} g^i(t,s)\,d\mu = C^i(0,S_0,\sigma_0^2),\ i = 1,\ldots,n, \quad \int_L e^{-rt} B_t\,d\mu = B_0 \\ &\mu : \mathcal{B}(L) \to [0,\infty) \text{ measure on the Borel sigma algebra } \mathcal{B}(L) \text{ on } L. \end{aligned} \tag{D_ϵ}$$

Based on the strong duality theorem we can deduce that (D_ϵ) has a solution and the optimal values $\nu(P_\epsilon)$ of (P_ϵ) and $\nu(D_\epsilon)$ of (D_ϵ) coincide. Note that $\int_L e^{-rt} B_t\,d\mu = B_0 \int_L d\mu = B_0\mu(L)$ such that the equality $B_0 = \int_L e^{-rt} B_t\,d\mu$ is equivalent to $\mu(L) = 1$. Thus the integral in (D_ϵ) is actually maximized over a set of probability measures.

Combining this observation with the statement of Theorem 1.15 shows that there exists a particular optimal solution of (D_ϵ) in the form of a discrete measure $\mu^* = \sum_{l=1}^{d} w_l \delta_{\{t_l,s_l\}}$, $d \in \mathbb{N}$, $d \leq n+1$, $0 \leq w_l \leq 1$, $\sum_{l=1}^{d} w_l = 1$, which is the weighted sum of d Dirac measures $\delta_{\{t_l,s_l\}}$ at the points $(t_l,s_l) \in L$ in which the constraints of (P_ϵ) are active, given the optimal solution α^*:

$$\alpha_0^* B_{t_l} + \sum_{i=1}^{n} \alpha_i^* g^i(t_l,s_l) = h_\epsilon(t_l,s_l), \quad l = 1,\ldots,d.$$

But (4.12) shows that the constraints of (P_ϵ) are not active $\forall\,(t,s) \in (T-\epsilon,T)\times\{D\}$ such that the points (t_l,s_l) defining the support of the Dirac measures must lie outside of this set. But on $L \setminus (T-\epsilon, T) \times \{D\}$ the functions h_ϵ and h coincide such that

$$\begin{aligned} \nu(D_\epsilon) &= \int_L e^{-rt} h_\epsilon(t,s) d\mu^*(t,s) = \sum_{l=1}^{d} e^{-rt_l} w_l h_\epsilon(t_l, s_l) \\ &= \sum_{l=1}^{d} e^{-rt_l} w_l h(t_l, s_l) = \int_L e^{-rt} h(t,s) d\mu^*(t,s). \end{aligned}$$

As μ^* is also feasible for (D), the latter equality proves that the optimal value $\nu(D)$ of (D) must be greater or equal to $\nu(D_\epsilon)$. But on the other hand the inequality $h(t,s) \leq h_\epsilon(t,s)$ on L implies that

$$\int_L e^{-rt} h(t,s) d\mu(t,s) \leq \int_L e^{-rt} h_\epsilon(t,s) d\mu(t,s) \quad \forall \mu,$$

and hence $\nu(D) \leq \nu(D_\epsilon)$ such that the optimal values of (D_ϵ) and (D) are equal and μ^* is also optimal for (D). Observing that the optimal values of (D_ϵ), (P_ϵ) and (P) all coincide finally proves the theorem. □

Theorem 4.8 shows that a suitable application of linear semi-infinite optimization theory delivers the dual of the static hedging problem along with a nice characterization of the feasible measures. Note that the proof of the theorem is based on a continuous modification of problem (4.10) instead of applying general convex duality theory based on the space of piecewise continuous functions (see Theorem 2.165 in Bonnans and Shapiro [13]). Despite of the problematic discontinuity the continuous modification allowed us to immediately apply standard results of the theory of continuous linear semi-infinite optimization, which in particular includes the existence of an optimal measure in the form of a finite sum of Dirac measures (see Hettich [52] or Bonnans and Shapiro [13], Proposition 5.104).

However, in the field of financial mathematics duality results are usually stated based on measures defined on the measurable space $(\Omega, \mathcal{F})$. Hence, to better compare Theorem 4.8 with duality results in the literature, the rest of this section is devoted to a restatement of problem (D) in the space $(\Omega, \mathcal{F})$.

As a first step we recall from Theorem 4.4 that problems (4.5), (P) are equivalent to the stochastic optimization problem

$$\begin{aligned} &\min_{\alpha \in \mathbb{R}^{n+1}} \alpha_0 B_0 + \sum_{i=1}^{n} \alpha_i C^i(0, S_0, \sigma_0^2) \\ &\text{s.t.} \quad \alpha_0 B_{\tau \wedge T} + \sum_{C^i \in \mathcal{C}, T_i \geq \tau \wedge T} \alpha_i C^i(\tau \wedge T, S_{\tau \wedge T}, \sigma_0^2) \geq C_{\text{uo}} \frac{B_{\tau \wedge T}}{B_T} = C_{\text{uo}}, \end{aligned} \tag{P_Ω}$$

where the latter equality holds since $C_{\text{uo}} = 0$ on $\{\tau < T\}$. Correspondingly, the optimal function values of problems (P), (P_Ω) and due to Theorem 4.8 also (D) coincide. This insight will be used in the following corollary to derive the dual of (P_Ω).

Corollary 4.9. *The dual of the static hedging problem* (P_Ω) *is given by*

$$\max_{\bar{Q}\in\mathbb{P}(\Omega)} \int_\Omega e^{-rT} C_{\mathrm{uo}}(\omega) d\bar{Q}(\omega)$$

$$\text{s.t.} \quad \int_\Omega e^{-r(\tau\wedge T_i)} C^i(\tau \wedge T_i, S_{\tau\wedge T_i}, \sigma_0^2) d\bar{Q}(\omega) = C^i(0, S_0, \sigma_0^2) \ \forall\, i, \qquad (\mathrm{D}_\Omega)$$

where $\mathbb{P}(\Omega)$ *denotes the set of all probability measures* $\bar{Q}$ *on* $(\Omega, \mathcal{F})$. *In particular problem* (D_Ω) *has a solution and the optimal values of* (P_Ω) *and* (D_Ω) *coincide.*

Moreover, there exists an optimal solution of the dual problem (D_Ω) *in the form of a discrete measure* $\bar{Q}^* = \sum_{l=1}^d x_l \delta_{\{\omega_l\}}$, $d \in \mathbb{N}$, $d \le n+1$, $0 \le x_l \le 1$, $\sum_{l=1}^d x_l = 1$, *which is the weighted sum of* d *Dirac measures* $\delta_{\{\omega_l\}}$ *at the points* $\omega_l \in \Omega$.

Proof. Let the optimal function values of (P), (P_Ω), (D) and (D_Ω) be denoted by $\nu(P)$, $\nu(P_\Omega)$, $\nu(D)$ and $\nu(D_\Omega)$, respectively. Based on Theorems 4.4, 4.8, we know that $\nu(D) = \nu(P) = \nu(P_\Omega)$. In the following we will show that $\nu(D_\Omega) \ge \nu(D)$ and $\nu(D_\Omega) \le \nu(P_\Omega)$ which proves that (D_Ω) is the dual of (P_Ω).

Due to Theorem 4.8 there exists a solution $\mu^* = \sum_{l=1}^d x_l \delta_{\{t_l, s_l\}}$ of (D) which is the finite sum of d Dirac measures. Let $\omega_l \in \{\tau\wedge T = t_l, S_{\tau\wedge T} = s_l\} \subset \Omega, l = 1, \dots, d$, and define $\bar{Q} := \sum_{l=1}^d x_l \delta_{\{\omega_l\}}$ as the finite sum of the Dirac measures $\delta_{\{\omega_l\}}$. Then

$$\begin{aligned}
&\int_\Omega e^{-r(\tau\wedge T_i)} C^i(\tau \wedge T_i, S_{\tau\wedge T_i}, \sigma_0^2) d\bar{Q}(\omega) \\
&= \sum_{l=1}^d x_l \int_\Omega e^{-r(\tau\wedge T_i)} C^i(\tau \wedge T_i, S_{\tau\wedge T_i}, \sigma_0^2) d\delta_{\{\omega_l\}} = \sum_{l=1}^d x_l e^{-rt_l} g^i(t_l, s_l) \\
&= C^i(0, S_0, \sigma_0^2),
\end{aligned}$$

where the second equality follows in analogy to the proof of Theorem 4.7 and the last equality is a consequence of the feasibility of μ^* for problem (D). Hence $\bar{Q}$ is feasible for (D_Ω) and its function value satisfies

$$\begin{aligned}
\int_\Omega e^{-rT} C_{\mathrm{uo}} d\bar{Q} &= \int_\Omega e^{-r(\tau\wedge T)} C_{\mathrm{uo}} d\bar{Q} \\
&= \sum_{l=1}^d x_l \int_\Omega e^{-r(\tau\wedge T)} C_{\mathrm{uo}} d\delta_{\{\omega_l\}} = \sum_{l=1}^d x_l e^{-rt_l} h(t_l, s_l) = \nu(D).
\end{aligned}$$

This implies $\nu(D_\Omega) \ge \nu(D)$. On the other hand let α^* be an optimal solution of (P),

(P_Ω). Then it follows that

$$\alpha_0^* B_{\tau\wedge T} + \sum_{C^i \in \mathcal{C}, T_i \geq \tau\wedge T} \alpha_i^* C^i(\tau\wedge T, S_{\tau\wedge T}, \sigma_0^2) \geq C_{\text{uo}} \quad (a.s.)$$

$$\iff e^{-r(\tau\wedge T)} \left(\alpha_0^* B_{\tau\wedge T} + \sum_{C^i \in \mathcal{C}, T_i \geq \tau\wedge T} \alpha_i^* C^i(\tau\wedge T, S_{\tau\wedge T}, \sigma_0^2) \right) \geq e^{-r(\tau\wedge T)} C_{\text{uo}}$$

$$\iff \alpha_0^* B_0 + \sum_{i=1}^{n} \alpha_i^* e^{-r(\tau\wedge T_i)} C^i(\tau\wedge T_i, S_{\tau\wedge T_i}, \sigma_0^2) \geq e^{-rT} C_{\text{uo}}.$$

Computing the expected values on both sides of this inequality with respect to a feasible measure $\bar{Q}$ of (D_Ω) leads to

$$\nu(P_\Omega) = \alpha_0^* B_0 + \sum\nolimits_{i=1}^{n} \alpha_i^* C^i(0, S_0, \sigma_0^2) \geq \int_\Omega e^{-rT} C_{\text{uo}} d\bar{Q}.$$

Since $\bar{Q}$ was arbitrary this implies $\nu(P_\Omega) \geq \nu(D_\Omega)$ and hence $\nu(P_\Omega) = \nu(D_\Omega)$. Finally, the optimal Dirac measure defined in the first part of the proof satisfies the properties in the second part of the theorem. □

Theorem 4.8 shows that the static super-replication problem is equivalent to an appropriate dual problem with a surprisingly simple characterization. In the dual problem, the expected discounted payoff of the barrier option is maximized over the set of all probability measures which are consistent with the prices of all calls in the static hedge portfolio at time $t = 0$. Formulation (D) of the dual problem reveals that restrictions on the feasible measures are only imposed along the crucial boundary L defining the barrier option payoff. Note that this property is also implicitly contained in problem (D_Ω) in the form of the random variables $\tau \wedge T_i$ and $S_{\tau\wedge T_i}$. However, apart from the crucial boundary L, the measures are undefined.

Compared with the case of dynamic super-replication, the dual problem looks very similar. In the dynamic case, the expected discounted payoff of the barrier option is maximized over the set of all equivalent martingale measures (see for example El-Karoui and Quenez [30]). Of course the class of equivalent martingale measures is much smaller than the set of measures consistent with the call prices at time $t = 0$, but the latter characterization is in line with the intuition of static hedging. The measures do not need to fulfill any dynamic (e.g. martingale) properties, but instead only have to provide a consistent static snapshot of the hedge portfolio.

The dual formulation of the static hedging problem also gives some advice about what is important to derive a successful static hedging strategy. As the feasible measures for the dual problem must be consistent with all model option prices, it is a natural requirement to ask the model option prices to fit the volatility surface observed in reality. Hence static hedge portfolios should be computed in models which calibrate well to the volatility surface. Stochastic volatility models are one possibility to achieve this goal, which is why we chose them as our general model setup.

As another unexpected insight Theorem 4.8 reveals that optimal model price consistent measures exist which are the weighted sum of at most $n + 1$ Dirac measures. In addition the support of those Dirac measures is exactly located at the points that are critical (active) in the sense of the primal problem. This means that such an optimal measure in the dual problem automatically puts weight on the points at which there is no safety cushion regarding hedge errors. Hence the identification of the active constraints of the primal problem is of practical as well as of theoretical importance. We will return to this idea in the next chapter which also contains an algorithm to numerically solve the semi-infinite optimization problem.

To summarize, the equivalent formulation of the static hedging problem as a linear semi-infinite optimization problem in Section 4.1 enabled us to derive the dual problem formulation in the space of probability measures. This formulation significantly contributes to the understanding and interpretation of the static hedging problem and allows to easily compare the concepts of static and dynamic super-replication. In the next chapter we will further exploit the semi-infinite structure to robustify the hedging problem and to set up algorithms to efficiently compute static trading strategies. In particular the derived algorithms will clearly outperform the Monte Carlo-based solution method presented in Section 3.2.

5 Eliminating Model Parameter Uncertainty

Even though the static hedging strategy derived in Chapter 3 already has attractive properties, the super-replication feature of the portfolio is based on the crucial assumption that the standard calls can be sold for the model prices C_τ^i at the time $\tau < T$ of a barrier hit. These prices in turn depend on the model parameters chosen at time $t = 0$, for example by calibrating the model to a given volatility surface. But as market data and hence implied model parameters change over time, the model and market prices of the calls might differ significantly at the future time of a barrier hit. Figure 5.1 illustrates for one example that a good fit of the volatility surface at time $t = 0$ can be arbitrarily bad in the future. This in turn may cause potentially huge hedging losses if the strategy computed in Chapter 3 is actually implemented in practice.

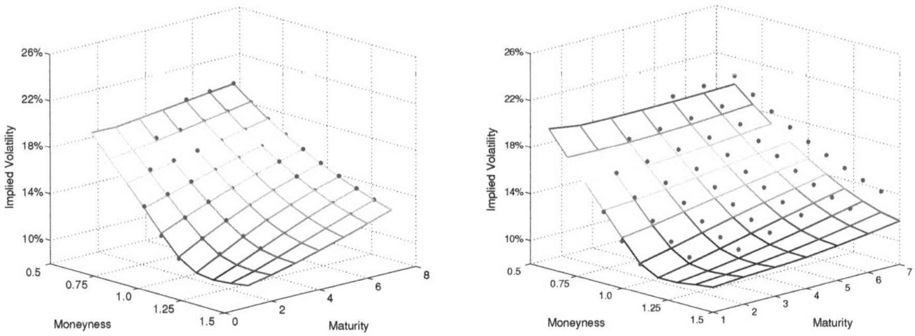

Figure 5.1: Market data (dots) and model fit for the Heston model at time $t = 0$ (left) and comparison of this fit with potentially changed market data in the future (right)

The negative effect of this so-called model parameter uncertainty was for example observed by Nalholm and Poulsen [72] for the static hedging strategies of Derman, Ergener and Kani [27] and Carr, Ellis and Gupta [24]. The authors show in a Monte Carlo simulation study, that the traditional static hedging approaches can result in extreme losses in comparison to the fair value of the barrier option if the model assumptions are subject to change.

To reduce this model parameter uncertainty, Allen and Padovani [2] propose a simple extension of the Derman et al. and Fink [38] approach (see Subsection 2.3.2) which additionally matches the value of the barrier option for a given finite number of shapes of the volatility surface at the time of the barrier hit. However, this extension suffers of the same problems as the Derman and Fink approach: Besides the fact that only a few volatility surface scenarios can be considered, the portfolio consists of a huge number of hedge instruments which can amount to several hundred standard options (depending on the number of scenarios). Furthermore the hedging problem is ill-posed, because one tries to exactly fit the discontinuous barrier option payoff by a linear combination of continuous call price functions. Correspondingly, the results of Allen and Padovani are very similar to the results of Fink and also show oscillations leading to extreme hedge errors in comparison to the fair value of the barrier option.

As an alternative approach to avoid model parameter uncertainty, Brown, Hobson and Rogers [19] derive upper and lower bounds for barrier options as well as a simple hedging strategy which is independent of the dynamics of the stock price and hence the financial market model (see Subsection 2.3.4). But by taking this extreme point of view the authors lose a lot of information and thus only obtain very conservative bounds for the price of a barrier option which are too rough for practical applications.

To summarize, model parameter uncertainty significantly affects static hedge portfolios for barrier options, but so far no method is known that allows to derive practical hedging strategies taking this uncertainty into account. In this chapter we present a new approach developed by Maruhn and Sachs in [70], [69], which makes use of the semi-infinite representation (see Section 4.1) of the static super-replication approach and incorporates the desired robustness into the description of the optimization problem. This can be achieved by parameter uncertainty sets in the sense of a worst case design and leads to a robust optimization problem with a strong nonlinearity in the constraint system. This nonlinearity turns out to cause the extreme model parameter uncertainty of static hedge portfolios for barrier options. However, by applying appropriate optimization methods we can control the nonlinearity to obtain truly robust static hedging strategies.

In a first step, Section 5.1 addresses model parameter uncertainty in the Black–Scholes model and hence robustifies static hedging strategies against volatility shocks. It turns out that the volatility risk can be eliminated by surprisingly low cost. Afterwards we move on to stochastic volatility models in Section 5.2 to quantify and eliminate the risk of changes of the whole curvature of the volatility surface (including the skew). Besides the theoretical insights of proving the existence of robust static hedging strategies, this chapter hence in detail analyzes the model parameter uncertainty (volatility and skew risk) for static hedge portfolios for barrier options and offers an intuitive solution to this problem.

Moreover we present algorithms to numerically solve the robust and non-robust static hedging problem by exploiting the linear semi-infinite representation of the original stochastic optimization problem. The complete elimination of the stochastic nature from the problem allows us to quickly solve even robustified versions of the hedging problem which would not be possible for the Monte Carlo-based solution procedure described in Section 3.2. Finally we prove convergence of the proposed algorithms to a solution of the robust static hedging problem in the continuous time financial market model.

5.1 Robust Static Hedging in the Black–Scholes Model

One of the key assumptions of the Black–Scholes model is that the volatility σ_0 at time $t = 0$ remains constant over time. In this section we will illustrate the effect of model parameter uncertainty in the sense of the volatility parameter and present the concept of how to robustify the static hedging strategy against changes of this parameter.

The starting point of our derivation is the semi-infinite formulation (4.5) of the super-replication problem for an up-and-out call with maturity T, strike K and barrier $D > K$ given by

$$\begin{aligned} &\min_{\alpha\in\mathbb{R}^{n+1}} \alpha_0 B_0 + \sum_{i=1}^{n} \alpha_i C^i(0, S_0, \sigma_0^2) \\ \text{s.t.} \quad &\alpha_0 B_t + \sum_{C^i\in\mathcal{C}, T_i\geq t} \alpha_i C^i(t, D, \sigma_0^2) \geq 0 \quad \forall\, t \in [0, T] \\ &\alpha_0 B_T + \sum_{C^i\in\mathcal{C}, T_i=T} \alpha_i (s - K_i)^+ \geq (s - K)^+ \quad \forall\, s \in [0, D]. \end{aligned} \tag{5.1}$$

It is well known (see (1.9)) that in the Black–Scholes model the call price $C^i(t, s) = C^i(t, s, \sigma^2)$, given a volatility $\sigma > 0$, is the solution of the parabolic differential equation

$$\begin{aligned} C_t + \tfrac{1}{2}\sigma^2 s^2 C_{ss} + (r - \delta) s C_s - rC &= 0 \quad && (t, s) \in (0, T_i) \times (0, \infty), \\ C(T_i, s) &= (s - K_i)^+ && s \in (0, \infty), \\ C(t, 0) &= 0 && t \in (0, T_i), \end{aligned} \tag{5.2}$$

which is solved by the Black–Scholes formula (4.6).

The section is structured as follows. In Subsection 5.1.1 we define the robust static hedging strategy and prove its existence under mild conditions. Afterwards Subsection 5.1.2 analyzes the structure of the semi-infinite optimization problem and describes the algorithm we use to solve the problem. In Subsection 5.1.3 we present a detailed numerical example, computing and comparing the non-robust and the robust solution given a set of liquidly exchange traded options. As it turns out, the robustness of the solution can be gained by relatively little additional cost such that the portfolio is attractive to option traders.

5.1.1 Description of the Robust Problem

To get an idea of how to robustify the static trading strategy against changes of the volatility parameter, we again take a look at the semi-infinite optimization problem (5.1). The constraints in this problem require a feasible strategy to super-replicate the barrier option at all possible times at which the barrier may be hit and for all attainable stock prices in case the barrier is not hit. In the case of no barrier hit the value of the remaining calls in the hedge portfolio with maturity $T_i = T$ only depends on the stock price s and hence is independent of the volatility. However, in case the barrier is hit at time $t < T$, the value of the calls $C^i(t, D, \sigma_0^2)$ by definition depends on the parameter σ_0.

To add robustness to the portfolio against volatility shocks it is hence a natural requirement to ask the super-replication property to hold for a whole set of volatility parameters instead of the single volatility σ_0. This intuitive idea is formalized in the sense of a worst case design in the next definition.

Definition 5.1. Let $\sigma_0 > 0$ be the implied volatility corresponding to current option prices in $t = 0$ and let $0 < \sigma_{\min} \leq \sigma_0 \leq \sigma_{\max}$ be a given volatility interval. A *cost-optimal robust super-replicating strategy* is defined as a solution of the robust optimization problem:

$$\begin{aligned}
&\min_{\alpha \in \mathbb{R}^{n+1}} \alpha_0 B_0 + \sum_{i=1}^{n} \alpha_i C^i(0, S_0, \sigma_0^2) \\
\text{s.t. } & \alpha_0 B_t + \sum_{C^i \in \mathcal{C}, T_i \geq t} \alpha_i C^i(t, D, \sigma^2) \geq 0 \;\; \forall\, (t, \sigma) \in [0, T] \times [\sigma_{\min}, \sigma_{\max}] \\
& \alpha_0 B_T + \sum_{C^i \in \mathcal{C}, T_i = T} \alpha_i (s - K_i)^+ \geq (s - K)^+, \;\; \forall\, s \in [0, D] \qquad (5.3) \\
& C^i(t, s, \sigma^2) = s e^{-\delta(T_i - t)} \Phi\left(\frac{\log(s/K_i) + (r - \delta + \sigma^2/2)(T_i - t)}{\sigma\sqrt{T_i - t}}\right) \\
& \qquad - K_i e^{-r(T_i - t)} \Phi\left(\frac{\log(s/K_i) + (r - \delta - \sigma^2/2)(T_i - t)}{\sigma\sqrt{T_i - t}}\right).
\end{aligned}$$

The feasible set of this optimization problem shall be denoted by $\mathcal{SR}(\sigma_{\min}, \sigma_{\max}) = \mathcal{SR}$. A strategy $\alpha \in \mathcal{SR}(\sigma_{\min}, \sigma_{\max})$ is called a robust super-replicating strategy.

Note that (5.3) is a linear semi-infinite optimization problem of the form

$$\begin{aligned}
&\min_{\alpha \in \mathbb{R}^{n+1}} c^{\mathrm{T}} \alpha \\
\text{s.t. } & a_1(t, \sigma)^{\mathrm{T}} \alpha \geq 0 \quad \forall\, (t, \sigma) \in [0, T] \times [\sigma_{\min}, \sigma_{\max}] \qquad (5.4) \\
& a_2(s)^{\mathrm{T}} \alpha \geq b_2(s) \quad \forall s \in [0, D],
\end{aligned}$$

where $a_1 : [0, T] \times [\sigma_{\min}, \sigma_{\max}] \to \mathbb{R}^{n+1}$ is a nonlinear map depending on the Black–Scholes formula and $a_2 : [0, D] \to \mathbb{R}^{n+1}$, $b_2 : [0, D] \to \mathbb{R}$ are continuous and piecewise linear.

Clearly, the feasible set $\mathcal{SR}(\sigma_{\min}, \sigma_{\max})$ of optimization problem (5.4) is closed and convex, but generally unbounded. However, the specific structure of the problem allows us to prove the existence of solutions.

Theorem 5.2. *If $\mathcal{SR}(\sigma_{\min}, \sigma_{\max})$ is nonempty, a solution of optimization problem (5.3) exists. Furthermore, the set of optimal solutions is convex and compact.*

Proof. Due to well-known theorems in linear semi-infinite optimization (see Theorem 1.12), it is sufficient to show that the objective function and $\mathcal{SR}(\sigma_{\min}, \sigma_{\max})$ have no direction of recession in common. Assume that $d \in \mathbb{R}^{n+1}$ is such a direction of recession. In the following we will show that d can only be equal to the zero vector.

Clearly, d is also a common direction of recession of the objective and the non-robust feasible set $\mathcal{SR}(\sigma_0, \sigma_0)$. This means that d satisfies

$$c^{\mathrm{T}} d \le 0, \quad a_1(t, \sigma_0)^{\mathrm{T}} d \ge 0 \ \forall t \in [0, T], \quad a_2(s)^{\mathrm{T}} d \ge 0 \ \forall s \in [0, D].$$

In analogy to the proof of Theorem 4.4 we can show that this set of constraints is equivalent to $\Pi_0(d) \le 0$, $\Pi_T(d) \ge 0$ (a.s.), where $\Pi_t(d)$ denotes the value of the static hedge portfolio d at time t.

If we assume that $\Pi_0(d) < 0$, then d would imply arbitrage opportunities which contradicts the fact that the Black–Scholes model is arbitrage-free in the set of static trading strategies. Hence $\Pi_0(d) = 0$ and $\Pi_T(d) \ge 0$ (a.s.). By the same argument $\Pi_T(d)$ must be equal to zero almost surely such that $\Pi_0(d) = 0$, $\Pi_T(d) = 0$ (a.s.). Again, an analogous argumentation as in the proof of Theorem 4.4 shows that these stochastic equalities are equivalent to the deterministic set of equations

$$c^{\mathrm{T}} d = 0, \quad a_1(t, \sigma_0)^{\mathrm{T}} d = 0 \ \forall t \in [0, T], \quad a_2(s)^{\mathrm{T}} d = 0 \ \forall s \in [0, D]. \tag{5.5}$$

As $a_1(t, \sigma_0)^{\mathrm{T}} d$ and $a_2(s)^{\mathrm{T}} d$ represent the value of the portfolio d, these equalities would imply that the payoffs of the bond and calls in the portfolio along the barrier and in case the barrier is not hit at all are linearly dependent. But the calls in the portfolio either have different strikes or maturities (see Assumption 4.2), hence it is intuitively clear that (5.5) can only hold for the trivial portfolio $d = 0$. However, the formal derivation of this result is not that simple and will cover the remainder of the proof.

By picking the call $C^j \in \mathcal{C}$ with maturity $T_j = T$ and the minimal strike $K_j := \min\{K_i : C^i \in \mathcal{C}, T_i = T\}$ we deduce from the last equality in (5.5) that

$$d_j(s - K_j) = -d_0 B_T \quad \forall\, s \in [K_j, K_j + \epsilon]$$

for some $\epsilon > 0$ such that $K_j + \epsilon < K_i$ for all $C^i \in \mathcal{C}$, $T_i = T, i \ne j$. As $B_T > 0$ does not depend on s, this can only hold true if $d_j = d_0 = 0$ such that (5.5) is independent of the bond position. Hence it is sufficient to prove that the payoff of the calls in the portfolio along the barrier and in case the barrier is not hit is linearly independent.

Without loss of generality we group the standard calls into sets $I_1, \dots, I_r$ with equal maturities $\bar{T}_1 < \bar{T}_2 < \dots < \bar{T}_r$. Then condition (5.5) together with $d_0 = 0$ implies

$$\sum_{i \in I_r} d_i C^i(t, D, \sigma_0^2) = 0 \ \forall\, t \in (\bar{T}_{r-1}, \bar{T}_r], \ \sum_{i \in I_r} d_i (s - K_i)^+ = 0 \ \forall\, s \in [0, D]. \tag{5.6}$$

By the superposition principle, the function $c(t,s) := \sum_{i\in I_r} d_i C^i(t,s,\sigma_0^2)$ also satisfies the Black–Scholes partial differential equation (5.2) with end condition $c(\bar{T}_r, s) = \sum_{i\in I_r} d_i (s-K_i)^+ \ \forall\, s \in (0,\infty)$.

Based on (5.2) and (5.6) we can conclude that $c(t,s)$ vanishes on $(\bar{T}_{r-1}, \bar{T}_r] \times \{0\}$, $(\bar{T}_{r-1}, \bar{T}_r] \times \{D\}$ and $\{\bar{T}_r\} \times [0,D]$. By the maximum principle, see e.g. Friedman [42], this implies $c(t,s) = 0$ on the strip $[\bar{T}_{r-1}, \bar{T}_r] \times [0,D]$. Since $c(t,D) = \frac{\partial c}{\partial s}(t,D) = 0$ on $(\bar{T}_{r-1}, \bar{T}_r]$, we can conclude by Mizohata's uniqueness theorem [71] that c vanishes everywhere on $[\bar{T}_{r-1}, \bar{T}_r] \times [0,\infty)$. In particular this implies $\sum_{i\in I_r} d_i(s-K_i)^+ = 0$ $\forall s \in (0,\infty)$ and hence $d_i = 0$ for $i \in I_r$.

This argument is repeated on the time strip $(\bar{T}_j, \bar{T}_{j+1}]$ with c defined analogously. By Assumption 4.2 we have $K_i \geq D$ for $i \in I_{j+1}$, $j+1 < r$ which guarantees $c(\bar{T}_{j+1}, s) = 0$ for $s \in [0,D]$. Hence, proceeding recursively, all coefficients d_i have to vanish which proves the theorem. □

Note that the assumption $\mathcal{SR}(\sigma_{\min}, \sigma_{\max}) \neq \emptyset$ in the theorem can easily be fulfilled by including a sufficiently large bond position or a call with the same strike and maturity as the up-and-out call in the portfolio. Hence the existence of a robust super-replicating strategy is assured for all cases of practical relevance. In the next subsection we will discuss the numerical solution of problem (5.3).

5.1.2 Numerical Solution

Optimization problem (5.3) is a linear semi-infinite programming problem where the parameters (t,σ) and s vary within compact sets. The numerical solution of this problem will in general depend on appropriate discretizations of these parameter sets. Let $M_1 \subset [0,T] \times [\sigma_{\min}, \sigma_{\max}]$, $|M_1| < \infty$, and $M_2 \subset [0,D]$, $|M_2| < \infty$, denote such discretizations, for example an equidistant grid in $[0,T] \times [\sigma_{\min}, \sigma_{\max}]$ and $[0,D]$, respectively. Then the discretized robust optimization problem is defined as

$$\begin{aligned} &\min_{\alpha\in\mathbb{R}^{n+1}} \ \alpha_0 B_0 + \sum_{C^i\in\mathcal{C}} \alpha_i C^i(0, S_0, \sigma_0^2) \\ \text{s.t.} \quad &\alpha_0 B_t + \sum_{C^i\in\mathcal{C}, T_i \geq t} \alpha_i C^i(t, D, \sigma^2) \geq 0 \ \ \forall\, (t,\sigma) \in M_1 \\ &\alpha_0 B_T + \sum_{C^i\in\mathcal{C}, T_i = T} \alpha_i (s-K_i)^+ \geq (s-K)^+ \ \ \forall\, s \in M_2. \end{aligned} \tag{5.7}$$

An algorithm will proceed by successively solving the subproblems (5.7), refining M_1, M_2 or exchanging points within these sets from one iteration to the next one. To guarantee convergence of such a procedure, it is crucial to verify the continuity of the constraints of the optimization problem with respect to their parameters $(t,\sigma) \in [0,T] \times [\sigma_{\min}, \sigma_{\max}]$ and $s \in [0,D]$. The following lemma states the corresponding result.

Lemma 5.3. *The semi-infinite constraint coefficients* $a_1 = (a_1^0, a_1^1, \dots, a_1^n)^{\mathrm{T}}$, $a_2 = (a_2^0, a_2^1, \dots, a_2^n)^{\mathrm{T}}$ *of optimization problem (5.3), (5.4) given by*

$$a_1^0(t,\sigma) = B_t, \quad a_1^i(t,\sigma) = \begin{cases} C^i(t, D, \sigma^2), & t \le T_i \\ 0, & t > T_i \end{cases}$$

$$a_2^0(s) = B_T, \quad a_2^i(s) = \begin{cases} (s - K_i)^+, & T_i = T \\ 0, & T_i < T, \end{cases}$$

as well as $b_2(s) = (s - K)^+$ *are continuous for all* $(t, \sigma) \in [0, T] \times [\sigma_{\min}, \sigma_{\max}]$ *and* $s \in [0, D]$, *respectively.*

Proof. For a_1^0, a_2^0, a_2^i, $i = 1, \dots, n$, and b_2 the continuity is obvious. The continuity of a_1^i, $i = 1, \dots, n$, immediately follows from the continuity of the mappings g^i in Lemma 4.6 with respect to t, and the Black–Scholes formula (4.6). □

As an implication of Lemma 5.3, problem (5.3) is a continuous linear semi-infinite programming problem with compact parameter sets. Applying the general theory of linear semi-infinite optimization, we can derive the following convergence theorem for solutions of the discretized problems.

Theorem 5.4. *Assume that* $\mathcal{SR}(\sigma_{\min}, \sigma_{\max}) \neq \emptyset$. *Then for each* $\epsilon > 0$ *there exist* $\delta_1, \delta_2 > 0$ *such that for each discrete subset* $M_1 \subset [0, T] \times [\sigma_{\min}, \sigma_{\max}]$, $|M_1| < \infty$ *and* $M_2 \subset [0, D]$, $|M_2| < \infty$ *satisfying*

$$\Delta(M_1) = \max_{(t,\sigma)\in[0,T]\times[\sigma_{\min},\sigma_{\max}]} \min_{(t_1,\sigma_1)\in M_1} \|(t,\sigma) - (t_1,\sigma_1)\|_2 \le \delta_1$$

$$\Delta(M_2) = \max_{s\in[0,D]} \min_{s_2\in M_2} |s - s_2| \le \delta_2$$

the discretized problem (5.7) is solvable and for each solution α_d^* *of (5.7) there exists a solution* α^* *of (5.3) such that* $\|\alpha_d^* - \alpha^*\|_2 \le \epsilon$.

Proof. Note that, as we have shown in the proof of Theorem 5.2, the feasible set $\mathcal{SR}(\sigma_{\min}, \sigma_{\max})$ and the objective have no direction of recession in common. Hence the theorem is an immediate consequence of linear semi-infinite optimization theory, see Theorem 1.13. □

Theorem 5.4 guarantees the convergence of solutions of the discretized problems (5.7) to solutions of the original problem (5.3) as we refine the grids M_1, M_2. Similar to Goberna and Lopez [46] we avoid unnecessary effort at each iteration by only refining the meshes locally in the neighborhood of nearly active constraints. This idea is the basis of Algorithm 2.

Note that due to Theorem 5.4 the existence of a solution of the discretized problem in step (S1) of the algorithm is guaranteed if the initial grids M_1, M_2 are sufficiently dense in the corresponding parameter sets.

Algorithm 2: Local Mesh-Refinement

Input: Let $M_1 \subset [0,T] \times [\sigma_{\min}, \sigma_{\max}]$, $|M_1| < \infty$ and $M_2 \subset [0,D]$, $|M_2| < \infty$ be given initial grids. Further let ϵ_1, ϵ_2 and TOL > 0 be given error tolerances. $d_1, d_2 > 1$ shall denote refinement parameters, let $k = 0$ be the initial iteration index.

Main Algorithm:

(S1) Calculate an optimal solution $\alpha^k \in \mathbb{R}^{n+1}$ of the discretized problem (5.7) with given sets M_1, M_2.

(S2) Calculate the slack at α^k:

$$\begin{aligned}\delta_1 &= \min_{(t,\sigma)\in[0,T]\times[\sigma_{\min},\sigma_{\max}]} a_1(t,\sigma)^{\mathrm{T}}\alpha^k, \\ \delta_2 &= \min_{s\in[0,D]} a_2(s)^{\mathrm{T}}\alpha^k - b_2(s)\end{aligned}$$

If $\min\{\delta_1, \delta_2\} \geq -\mathrm{TOL}$ then stop.

(S3) For each $(t,\sigma) \in M_1$, $s \in M_2$ Do

If $a_1(t,\sigma)^{\mathrm{T}}\alpha^k < \epsilon_1$ locally refine M_1 around (t,σ).
If $a_2(s)^{\mathrm{T}}\alpha^k - b_2(s) < \epsilon_2$ locally refine M_2 around s.

End Do
Set $k \leftarrow k+1$, $\epsilon_1 \leftarrow \epsilon_1/d_1$, $\epsilon_2 \leftarrow \epsilon_2/d_2$ and go to step (S1).

Furthermore, we can prove in analogy to Lemma 5.3 that a_1^i is twice continuously differentiable in case $K_i > D$. If $K_i = D$ the smoothness is preserved on $([0,T_i) \cup (T_i,T]) \times [\sigma_{\min}, \sigma_{\max}]$, but a corner occurs at $\{T_i\} \times [\sigma_{\min}, \sigma_{\max}]$. However, these corners do not pose a problem, because they are known in advance and hence we can solve the minimization problem in step (S2) on appropriate subintervals of $[0,T] \times [\sigma_{\min}, \sigma_{\max}]$ by a Newton-based method. Of course, the points resulting from this optimization can also be included in the sets M_1, M_2 for the next iteration.

Regarding the mesh refinement in step (S3), we decided to add the following points to the sets M_1, M_2:

$$\begin{aligned}M_1 &\leftarrow M_1 \cup \left\{ \begin{pmatrix} t-\Delta t \\ \sigma-\Delta\sigma \end{pmatrix}, \begin{pmatrix} t-\Delta t \\ \sigma+\Delta\sigma \end{pmatrix}, \begin{pmatrix} t+\Delta t \\ \sigma-\Delta\sigma \end{pmatrix}, \begin{pmatrix} t+\Delta t \\ \sigma+\Delta\sigma \end{pmatrix} \right\} \\ M_2 &\leftarrow M_2 \cup \{s-\Delta s, s+\Delta s\},\end{aligned}$$

where Δt, $\Delta\sigma$ and Δs are successively reduced during the iteration process.

Now we illustrate the efficiency of the proposed method and the obtained hedge portfolios by several numerical examples.

5.1.3 An Example of a Hedge Portfolio

In this subsection we solve optimization problem (5.3) with the proposed algorithm to compute static hedging strategies with varying degree of robustness. Following the Black–Scholes example in Subsection 3.3.1, our goal is to hedge an up-and-out call with strike $K = 2750$, barrier $D = 3300$ and maturity $T = 1$. As the underlying of the barrier option we choose the EURO STOXX 50 index with price $S_0 = 2750$ in September 2004. Furthermore, the risk-free rate and the dividend yield are assumed to satisfy $r = 5.5\%$ and $\delta = 2.5\%$. The implied volatility at time $t = 0$ shall be given by $\sigma_0 = 20\%$.

For these parameters, the value of the up-and-out call C_{uo} can be calculated via the following closed form formula (see for instance [89]) to be approximately $\pi(C_{uo}) = 1.13\%$ of the spot S_0 at time $t = 0$:

$$\begin{aligned}\pi(C_{uo}) &= \mathrm{Call}(S_0, K) - \mathrm{Call}(S_0, D) - (D-K)e^{-rT}\Phi(d_1) - \\ &\quad -\left(\frac{D}{S_0}\right)^{\frac{2\nu}{\sigma^2}} \left\{\mathrm{Call}\left(\frac{D^2}{S_0}, K\right) - \mathrm{Call}\left(\frac{D^2}{S_0}, D\right) - (D-K)e^{-rT}\Phi(d_2)\right\} \\ \nu &= r - \delta - \frac{\sigma^2}{2}, \qquad d_1 = \frac{\log\left(\frac{S_0}{D}\right) + \nu T}{\sigma\sqrt{T}}, \qquad d_2 = \frac{\log\left(\frac{D}{S_0}\right) + \nu T}{\sigma\sqrt{T}}.\end{aligned}$$

Here $\mathrm{Call}(u, v)$ denotes the Black–Scholes value (4.6) of a standard call at time $t = 0$ with strike v, maturity $T = 1$, value of the underlying u at $t = 0$ and $\sigma_0 = 20\%$. Further $\Phi(\cdot)$ is the cumulative normal distribution.

To compute a static hedging strategy for the up-and-out call, we have to decide which calls C^i should be included in the hedge portfolio. As in Section 3.3, we choose C^i to be the EURO STOXX 50 standard calls listed in Table 5.1. Note that the strikes and maturities satisfy Assumption 4.2. In addition the call with maturity 1.00 and strike 2750 guarantees that a super-replicating static strategy exists - that means the feasible set $\mathcal{SR}(\sigma_{\min}, \sigma_{\max})$ is nonempty for any choice of $0 < \sigma_{\min} \le \sigma_0 \le \sigma_{\max}$. Hence, by Theorem 5.2 the linear semi-infinite optimization problem (5.3) has a solution.

	C^1	C^2	C^3	C^4	C^5	C^6	C^7	C^8	C^9	C^{10}
T_i	1.00	1.00	1.00	1.00	1.00	0.75	0.75	0.75	0.50	0.50
K_i	2750	3300	3350	3450	3600	3300	3400	3600	3300	3500

Table 5.1: Standard calls C^i included in the hedge portfolio

To illustrate the behavior of the semi-infinite optimization Algorithm 2 and to compare it with the stochastic solution method applied in Section 3.3, we first compute the non-robust static hedge portfolio ($\sigma_{\min} = \sigma_0 = \sigma_{\max}$). After numerically testing the portfolio sensitivity with respect to the volatility parameter σ, we increase the degree of robustness by gradually extending the robustness interval $[\sigma_{\min}, \sigma_{\max}]$.

Solution of the Non-Robust Problem To solve the non-robust optimization problem, we choose the following parameters in Algorithm 2: $\epsilon_1 = 10^{-1}$, $d_1 = 10$ and a tolerance of TOL $= 10^{-8}$ for the hedge error in percent of the underlying S_0. For the non-robust problem the compact set $[0, T] \times [\sigma_{\min}, \sigma_{\max}]$ is in fact just the line $[0, T] \times \{\sigma_0\}$, hence it suffices to discretize the time interval. We start with an initial grid M_1 consisting of 41 time steps.

A quick analysis[1] shows that by neglecting the inefficient bond position ($\alpha_0 = 0$) we can replace the semi-infinite constraint $a_2(s)^{\mathrm{T}}\alpha \geq b_2(s)\ \forall\ s \in [0, D]$ by the single constraint $a_2(D)^{\mathrm{T}}\alpha \geq b_2(D)$ due to the particular choice of the calls in Table 5.1. Hence we can omit the grid M_2 as well as the parameters ϵ_2 and d_2 during the optimization.

Table 5.2 presents the results from the iteration process. Here the cost of the hedge portfolio at iteration k, denoted by $\Pi_0(\alpha^k)$, as well as the slack $\min\{\delta_1, \delta_2\}$ are listed in percent of the underlying S_0. Obviously, Algorithm 2 successfully terminates after five outer iterations with the optimal solution given in Table 5.3, reaching a slack of less than 10^{-8}. The cost of the hedge portfolio converges to an optimal value of 1.14%, which is only marginally more expensive than the barrier option itself ($\pi(C_{\mathrm{uo}}) =$ 1.13%).

Iteration	Slack	$\Pi_0(\alpha^k)$	$\lvert M_1 \rvert$	ϵ_1
0	−2.069330e−004	1.1408456e−002	41	—
1	−4.875932e−005	1.1414368e−002	68	1.00000e−001
2	−3.610492e−006	1.1414963e−002	95	1.00000e−002
3	−3.981831e−007	1.1415065e−002	122	1.00000e−003
4	−4.422266e−008	1.1415076e−002	149	1.00000e−004
5	−4.851917e−009	1.1415077e−002	208	1.00000e−004

Table 5.2: Iteration process for the non-robust problem

	C^1	C^2	C^3	C^4	C^5	C^6	C^7	C^8	C^9	C^{10}
T_i	1.00	1.00	1.00	1.00	1.00	0.75	0.75	0.75	0.50	0.50
K_i	2750	3300	3350	3450	3600	3300	3400	3600	3300	3500
α_i	1.00	−33.02	40.06	−8.20	0.79	−0.04	0.04	−0.25	0.02	−0.09

Table 5.3: Optimal portfolio weights α_i for the non-robust problem

Comparing Table 5.3 with Table 3.1 immediately reveals that the optimal solution of the deterministic semi-infinite optimization problem computed with Algorithm 2 and the solution of the stochastic optimization problem derived with Algorithm 1 coincide.

[1] The only call satisfying $T_i = T$ and $K_i < D$ is the standard call C^1 with maturity $T_1 = 1.00$ and strike $K_1 = 2750 = K$. Hence the linear system $a_2(s)^{\mathrm{T}}\alpha = \alpha_1(s - 2750)^+ \geq b_2(s) = (s - 2750)^+$ is linearly dependent and reduces to the single constraint $\alpha_1(D - 2750)^+ \geq (D - 2750)^+$ which is equivalent to $\alpha_1 \geq 1$.

Of course this is not surprising since we proved the equivalence of the two problems in Theorem 4.4. However, the Monte Carlo-based solution method requires the simulation of M sample paths of the stock price, where M usually has to be chosen quite large. In contrast to this, the solution of the linear semi-infinite optimization problem only takes a fraction of the time it takes to solve the stochastic optimization problem. In fact, the computation time was boiled down from several minutes to less than five seconds on a usual PC with 3 GHz CPU.

The reason for this speed-up is twofold. Firstly, the semi-infinite problem representation completely avoids the numerical simulation of stochastic differential equations. Secondly, the deterministic method automatically refines the mesh in relevant areas whereas the stochastic simulation method randomly generates mesh points and hence takes a very long time to cover the whole time interval $[0, T]$. Based on the adaptive method shown in Table 5.2, the number of grid points $|M_1|$ only increases by a factor of five until sufficient accuracy is reached. This mesh refinement is graphically visible in Figure 5.2, which plots the grid M_1 and the hedge error of the optimal portfolio for all times $t \in [0, T]$ in case of a barrier hit.

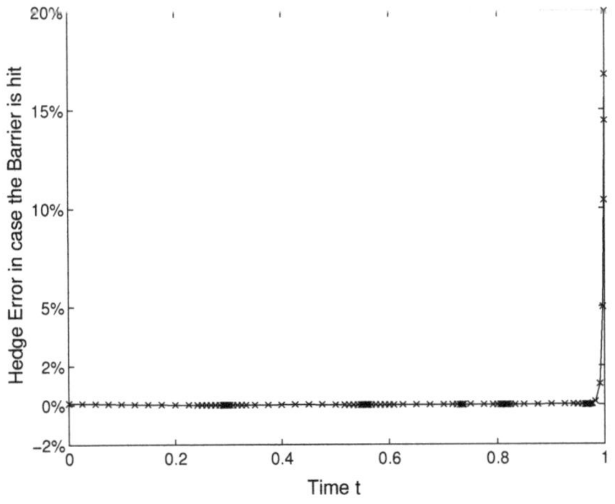

Figure 5.2: Hedge error $a_1(t, \sigma_0)^{\mathrm{T}}\alpha$ on the barrier for the optimal portfolio. Points marked by an x belong to the adaptively refined grid M_1

Obviously, Algorithm 2 first covered the interval $[0, T]$ with a rough grid, and then successively refined the mesh M_1 in the vicinity of nearly active constraints. In the limit, the method hence precisely identifies the active constraints which also form the support of the optimal Dirac measure solving the dual problem (see Theorem 4.8). From a financial point of view, the active constraints are the points in time, in which a trader has to pay particular attention to prevent losses in case the barrier is hit. These crucial points are the ones the proposed method automatically steers at.

As mentioned before, the key assumption of the hedge portfolio derived so far is, that the portfolio can be sold anytime the barrier is hit for model prices based on the volatility σ_0 chosen at $t = 0$. If the volatility at the hitting time differs from the initial volatility, the hedge portfolio might not offer the protection expected from a super-replication portfolio. To analyze this, we simulated $M = 100\,000$ sample paths[2] of the Black–Scholes stochastic differential equation to determine the simulated hedge error for different volatility states σ. In particular we focus on the loss the bank might encounter by selling the barrier option and buying the hedge portfolio consisting of standard calls. The results are shown in Table 5.4. The average and largest loss are listed in percent of the underlying S_0.

	5%	10%	15%	20%	25%	30%
Average Loss	−0.003%	−0.105%	−0.102%	0.000%	0.000%	0.000%
Largest Loss	−4.657%	−1.540%	−0.608%	0.000%	0.000%	0.000%
Probability of Loss	0.10%	9.35%	25.01%	0.00%	0.00%	0.00%

Table 5.4: Simulation results for the non-robust optimal portfolio and various volatility states σ of the Black–Scholes model. The results are based on 100 000 sample paths of the underlying.

Obviously, the non-robust super-replication portfolio offers protection for the set of volatility states $\sigma \geq \sigma_0 = 20\%$. However, as the volatility gradually decreases below 20%, large hedge errors may occur for a significant number of future states of the market. Although the probability of hitting the barrier and encountering portfolio losses decreases as we reduce the volatility, a probability of approximately 9% for quite large losses signalizes a problem of the hedge portfolio. Moreover, compared to the price of the barrier option $\pi(C_{uo}) = 1.13\%$, a largest loss of for example -1.54% in case $\sigma = 10\%$ is unacceptable and contradicts the notion of a super-replication strategy.

Figure 5.3 gives some further insight into the structure of the error, showing a continuous plot of the hedge error on the barrier for volatilities ranging from 0% to 40%. Although it looks on first sight as if the characteristics of the hedge error plotted in Figure 5.2 carry over to the other volatility states, a closer look at the minimal hedge error over time reveals that the portfolio is exactly fitted to the volatility state $\sigma = 20\%$. An increase in volatility also increases the hedge error, but does not lead to a loss. In contrast to this, a decreasing volatility may lead to significant losses which can be multiples of the initial value of the barrier option.

In summary we can conclude that the optimal non-robust hedge portfolio, although one is tempted to assume a general super-replication property, only offers protection for volatility states greater or equal to the initial volatility $\sigma_0 = 20\%$. Volatility shocks below this initial volatility may result in large hedging losses which are undesired in practice. To avoid these losses, we have to compute an optimal solution which is robust

[2]For the time discretization of the stochastic differential equation, we used the Euler-Maruyama Scheme with step size $\Delta t = 0.001$.

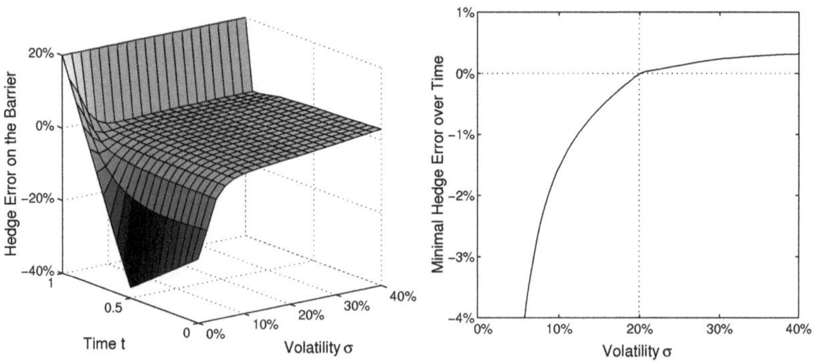

Figure 5.3: Hedge error on the barrier for the non-robust optimal portfolio and a wide variety of volatility states

with respect to changes in the volatility.

Adding Robustness to the Solution Having experienced the sensitivity of the previously computed hedge portfolio with respect to changes of the model parameter σ, we now aim at adding robustness to the super replication strategy. Mathematically, we will solve optimization problem (5.3) with $\sigma_{\min} < \sigma_0 = 20\% < \sigma_{\max}$. From a financial point of view, this guarantees the super-replication property for volatility states $\sigma \in [\sigma_{\min}, \sigma_{\max}]$. We begin by demanding robustness for the interval $[15\%, 25\%]$, hence $\sigma_{\min} = 15\%$, $\sigma_{\max} = 25\%$.

To solve the robust optimization problem (5.3) we choose the following parameters in Algorithm 2: $\epsilon_1 = 1.0$, $d_1 = 5$ and $\text{TOL} = 10^{-8}$. In contrast to the non-robust problem, the set $[0, T] \times [\sigma_{\min}, \sigma_{\max}]$ is now two-dimensional, such that we have to discretize in both the time- and the volatility-dimension. We start with an equidistant grid M_1, consisting of a total of $21 \cdot 21 = 441$ points. By the same argument as in the non-robust case, we can omit M_2, ϵ_2 and d_2 from the optimization.

The output of Algorithm 2 and the corresponding optimal solution are listed in Tables 5.5 and 5.6, respectively. Again, the slack as well as the cost $\Pi_0(\alpha^k)$ of the portfolio are expressed in percent of the underlying S_0.

The algorithm terminates after eight iterations with numerically zero slack. The optimal portfolio has a cost of 1.19% which is only slightly more expensive than the non-robust portfolio (1.14%) and the barrier option (1.13%). However, in contrast to the non-robust portfolio, the robust solution offers protection for the desired volatility range $[15\%, 25\%]$ as can be seen in Figure 5.4. The graph of the hedge error also shows the local mesh refinement around the nearly active constraints. During the iteration process, the number of grid points increases approximately by a factor of three.

Figure 5.4 also reveals the points of non-differentiability of the hedge-error function on the barrier. Recall from the end of Subsection 5.1.2 that these points are given by

Iteration	Slack	$\Pi_0(\alpha^k)$	$\lvert M_1 \rvert$	ϵ_1
0	−1.604109e−004	1.185534e−002	441	—
1	−1.120923e−004	1.187441e−002	650	1.00000e+000
2	−2.979500e−005	1.188517e−002	745	2.00000e−001
3	−5.391230e−005	1.188703e−002	823	4.00000e−002
4	−7.187315e−005	1.188732e−002	891	8.00000e−003
5	−9.429134e−006	1.188748e−002	967	1.60000e−003
6	−6.949047e−005	1.188753e−002	1035	3.20000e−004
7	−6.914103e−007	1.188755e−002	1103	6.40000e−005
8	0.000000e+000	1.188756e−002	1171	1.28000e−005

Table 5.5: Iteration process for the robust problem

	C^1	C^2	C^3	C^4	C^5	C^6	C^7	C^8	C^9	C^{10}
T_i	1.00	1.00	1.00	1.00	1.00	0.75	0.75	0.75	0.50	0.50
K_i	2750	3300	3350	3450	3600	3300	3400	3600	3300	3500
α_i	1.00	−30.87	35.66	−4.31	−1.46	0.01	0.15	−0.10	0.00	0.06

Table 5.6: Optimal portfolio weights α_i for the robust problem

$\{T_i\}\times[\sigma_{\min}, \sigma_{\max}]$ for calls C^i with $K_i = D$ and $T_i < T$. In our case, these are the two calls C^6 and C^9 leading to non-differentiabilities along the lines $\{0.5\} \times [15\%, 25\%]$ and $\{0.75\} \times [15\%, 25\%]$. However, as we can optimize on the smooth subintervals, the non-differentiabilities do not pose a problem.

Compared to the non-robust solution, the qualitative behavior of the hedge observed in Figure 5.2 now carries over to the whole volatility interval. The robust portfolio has a relatively small hedge error for small t and offers a comfortable safety cushion close to maturity. The price for this additional protection, preventing possibly large hedging losses, seems to be quite low. In comparison to the non-robust portfolio, the price only increases by five basis points.

Of course the obtained robust hedge portfolio might still deteriorate if the volatility drops below $\sigma_{\min} = 15\%$. This is clearly visible in Figure 5.5 which shows the minimal hedge error for the robust portfolio. However, the probability that the portfolio fails to hedge the barrier option is much lower than in the non-robust case.

To offer protection for the additional volatility states not covered by the robust hedge portfolio derived so far, we have to increase the volatility interval within the robust optimization problem (5.3). Intuitively, the cost of a robust portfolio will increase as we enhance the protection to a larger volatility interval. Hence the question of how large to choose the volatility interval is merely a question of how risk-averse the portfolio trader is. By applying utility theory, there will be an optimal risk/cost-combination for every trader.

As examples for the possible degrees of robustness and the associated cost-effects we list several risk/cost combinations in Table 5.7. For instance a classical risk-averter

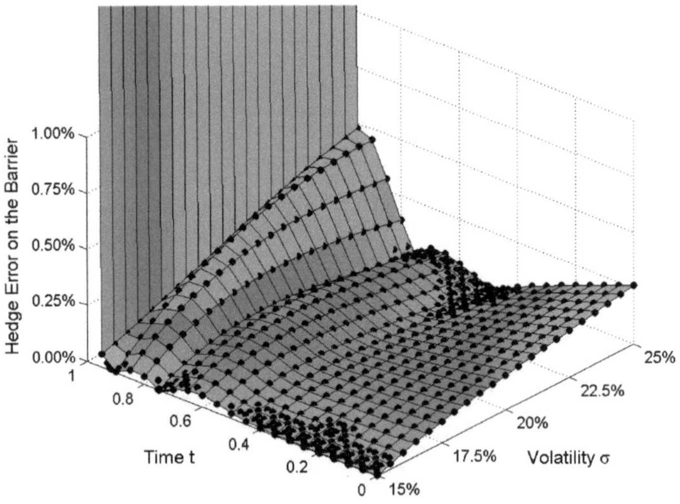

Figure 5.4: Hedge error on the barrier for the robust optimal portfolio and a wide variety of volatility states

would choose the robust portfolio offering protection for the set of volatility states $[0\%, 100\%]$ with an additional cost of about 34 basis points in comparison to the non-robust portfolio. This risk premium is surprisingly low for a robust portfolio also covering extreme cases. Between these extreme scenarios of the non-robust and the risk-averter portfolio the portfolio cost grows linearly in the size of the volatility uncertainty set $[\sigma_{\min}, \sigma_{\max}]$.

$\sigma_{\min}$	20%	15%	10%	5%	0%
$\sigma_{\max}$	20%	25%	30%	50%	100%
Cost	1.14%	1.19%	1.27%	1.34%	1.48%

Table 5.7: Cost of robust hedge portfolios with varying degree of robustness

The presented examples show that the semi-infinite formulation of the static super-replication problem allows to efficiently robustify the hedging approach against volatility shocks. From the theoretical side, the results are underlined with an existence proof of an optimal solution and the convergence of solutions of the discretized problem to a hedging strategy in the continuous time financial market model. Furthermore, the proposed algorithm quickly solves the semi-infinite problem and hence is superior to the Monte Carlo-based approach used in Chapter 3 to solve the equivalent stochastic optimization problem.

After we successfully robustified the static hedge portfolio against volatility shocks, the question remains if these results transfer in analogy to changes in the curvature of the volatility surface. This will be the focus of the next section.

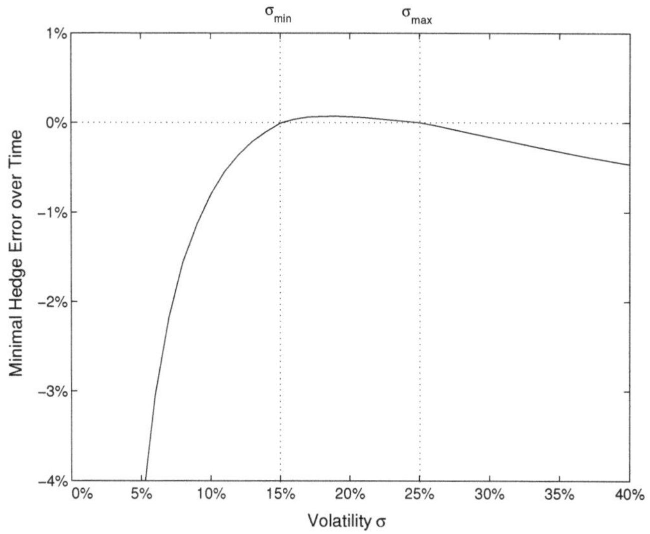

Figure 5.5: Minimal hedge error on the barrier for the robust optimal portfolio

5.2 Robust Static Hedging in Stochastic Volatility Models

By definition, static hedge portfolios for barrier options consist of standard calls with different strikes and maturities which cover a broad range of the volatility surface. For the case of the Black–Scholes model, the previous section analyzed and eliminated the sensitivity of such a portfolio with respect to volatility shocks. But in this model the volatility surface is flat such that a robustification against volatility shocks guarantees the super-replication property if the initial (flat) volatility surface moves up or down. However, in reality the volatility surface is not flat. Instead it has a specific curvature (in case of equity indices typically a skew) which changes over time. This change can consist of simple variations in height (volatility shocks), but also of very complicated changes of the whole curvature of the surface. In particular the volatility surface at time $t = 0$ can differ significantly from the one at the time of the barrier hit. Correspondingly, we have to additionally robustify the static super-replication portfolio against changes in the curvature of the volatility surface.

The starting point of our analysis has to be a financial market model which usually provides a good fit of the volatility surface (or at least the part of the surface containing the calls in the portfolio). For this purpose the general stochastic volatility model stated in Assumption 4.1 is sufficient. In this model Theorem 4.4 showed that the stochastic static super-replication problem (4.3) is equivalent to the deterministic semi-infinite optimization problem

$$
\begin{aligned}
&\min_{\alpha\in\mathbb{R}^{n+1}} \ \alpha_0 B_0 + \sum_{i=1}^{n} \alpha_i C^i(0, S_0, Y_0) \\
\text{s.t.} \quad & \alpha_0 B_t + \sum_{C^i\in\mathfrak{C}, T_i\geq t} \alpha_i C^i(t, D, y) \geq 0 \quad \forall\, (t, y) \in \Theta_1 \\
& \alpha_0 B_T + \sum_{C^i\in\mathfrak{C}, T_i = T} \alpha_i (s - K_i)^+ \geq (s - K)^+ \quad \forall\, s \in \Theta_2,
\end{aligned}
\tag{5.8}
$$

where Θ_1 and Θ_2 are given by $\Theta_1 = \operatorname{supp}\big(Q^{(\tau, Y_{\tau\wedge T})}\big) \cap \big([0, T] \times \bar{\mathcal{Y}}\big)$ and $\Theta_2 = \{s \in [0, D] : (\infty, s) \in \operatorname{supp}(Q^{(\tau, S_T)})\}$.

In Subsection 5.2.1 we define a robustified version of problem (5.8) also taking skew-risk into account. After proving the existence of an optimal solution, we present an algorithm to numerically solve the optimization problem in Subsection 5.2.2. As it turns out, the curse of dimensionality prevents a local mesh refinement in the form of Algorithm 2. Instead we adapt an exchange method to the static hedging problem which successively identifies the worst case constraints and adds them to the parameter grid. Finally, Subsection 5.2.3 presents a detailed numerical example. In this example, we exactly quantify the skew risk and compute the cost to eliminate this risk from the hedge portfolio. Although the skew risk turns out to be significant, the additional robustness can be gained by surprisingly low cost.

5.2.1 Definition of the Robust Problem

To derive a robust static hedging strategy we have to guarantee that the value of the hedge portfolio is always greater or equal to the value of the up-and-out call. The semi-infinite formulation (5.8) of the static hedging problem reveals that we have to consider the two separate cases of a barrier hit and no barrier hit before maturity T.

In case the barrier is not hit from $t = 0$ to $t = T$, the super-replication property requires the calls with maturity T to provide a payoff greater or equal to the payoff of the up-and-out call for all possible stock prices S_T that can be attained without crossing or touching the barrier. This is reflected by the constraints

$$
\alpha_0 B_T + \sum_{C^i\in\mathfrak{C}, T_i = T} \alpha_i (s - K_i)^+ \geq (s - K)^+ \quad \forall\, s \in \Theta_2,
$$

where $\Theta_2 = [0, D]$ for most models of practical interest (see Example 4.5). By definition, these constraints only depend on the stock price and hence are not affected by movements of the volatility surface.

The cases of a barrier hit are reflected by the first constraints in optimization problem (5.8) guaranteeing that the calls in the portfolio provide a payoff greater or equal to zero

$$\alpha_0 B_t + \sum_{C^i \in \mathfrak{C}, T_i \geq t} \alpha_i C^i(t, D, y) \geq 0 \quad \forall\, (t, y) \in \Theta_1.$$

Note that the value of the calls $C^i(t, D, y)$ at time t of the barrier hit by definition depends on the market prices and hence the curvature of the volatility surface at that time. In the usual stochastic volatility models this curvature is parametrized by the short term variance y, the correlation ρ of the Brownian motions and some additional model parameters $x \in \mathbb{R}^k$. At time $t = 0$ the model is calibrated to the current market data leading to implied parameters Y_0, ρ_0 and x_0. But while the stochastic volatility model by definition allows the short term variance Y_t at time t of a barrier hit to differ from Y_0, the other model parameters are assumed to remain constant. In terms of the hedge portfolio this means that the super-replication constraints above only guarantee protection against movements of the short term variance y at the time of a barrier hit, but not against moves of the other surface parameters ρ and x. Hence the performance of the static hedge portfolio considered so far might deteriorate rapidly if the additional skew parameters change over time.

Observing this, it is an intuitive idea to ask the super-replication property in case of a barrier hit to further hold for a whole uncertainty set $U \subset \mathbb{R}^{k+1}$ of model parameters $p = (x^{\mathrm{T}}, \rho)^{\mathrm{T}}$. This leads to the following definition of a robust static hedging strategy.

Definition 5.5. Suppose that the assumptions of Theorem 4.4 are satisfied such that the problem (4.3) of finding the cost-optimal static super-replication strategy is equivalent to problem (5.8). Further assume that the functions β, γ and σ describing the financial market model (4.1) depend on model parameters $x \in \mathbb{R}^k$ such that $\beta = \beta(y, x)$, $\gamma = \gamma(y, x)$ and $\sigma = \sigma(y, x)$. Accordingly the call prices $C^i(t, D, y)$ also depend on x and in addition the correlation coefficient ρ of the Brownian motions such that $C^i(t, D, y) = C^i(t, D, y, p_0)$ with $p_0 := (x^{\mathrm{T}}, \rho)^{\mathrm{T}} \in \mathbb{R}^{k+1}$. Let $U \subset \mathbb{R}^{k+1}$ be a compact set of model parameters with $p_0 \in U$. In addition we impose the assumption that for each parameter $p = (\tilde{x}^{\mathrm{T}}, \tilde{\rho})^{\mathrm{T}} \in U$ the stochastic differential equations (4.1) with coefficients $\beta(y, \tilde{x}), \gamma(y, \tilde{x}), \sigma(y, \tilde{x})$ and correlation parameter $\tilde{\rho}$ are well defined with $Y_t = Y_t(p) \in \mathcal{Y}$ (a.s.). Let the price of a call option in the resulting financial market model be denoted by $C^i(t, s, y, p)$ and assume that the sets Θ_1, Θ_2 in Theorem 4.4 are independent of $p \in U$. Then a "cost-optimal robust static super-replication strategy" is defined as a solution of the robust optimization problem

$$\begin{aligned} &\min_{\alpha \in \mathbb{R}^{n+1}} \alpha_0 B_0 + \sum_{i=1}^{n} \alpha_i C^i(0, S_0, Y_0, \mathbf{p_0}) \\ \text{s.t.} \quad &\alpha_0 B_t + \sum_{C^i \in \mathfrak{C}, T_i \geq t} \alpha_i C^i(t, D, y, \mathbf{p}) \geq 0 \quad \forall\, (t, y, \mathbf{p}) \in \Theta_1 \times \mathbf{U} \qquad (5.9) \\ &\alpha_0 B_T + \sum_{C^i \in \mathfrak{C}, T_i = T} \alpha_i (s - K_i)^+ \geq (s - K)^+ \quad \forall\, s \in \Theta_2. \end{aligned}$$

Clearly, the robust optimization problem (5.9) is a natural extension of the non-robust problem (5.8) taking the model parameter uncertainty into account in the sense of a worst case design. While problem (5.8) only requires the super-replication property to hold for the model parameter p_0 chosen at time $t = 0$, the robust static hedging strategy defined above guarantees this property for a whole set of model parameters $p \in U$. From a financial point of view this means that the hedge portfolio super-replicates the up-and-out call for all possible future market prices which correspond to or are close to volatility surfaces associated with some model parameter $p \in U$.

Although the robust optimization problem (5.9) has attractive properties, it is still unclear if a solution of this problem exists at all. In the remainder of this subsection we will prove the existence of a solution under the following mild conditions.

Assumption 5.6. Assume that the coefficient functions β, γ and σ of financial market model (4.1) are continuous. Further assume that there exists an open set $\mathbb{O} \subset \mathcal{Y}$ such that β, γ and σ are infinitely often differentiable on $\mathbb{O}$ and $\sigma(y), \gamma(y) \neq 0$ for all $y \in \mathbb{O}$. Let the support of the joint distribution of (S_t, Y_t) under the equivalent martingale measure Q be equal to $[0, \infty) \times \bar{\mathcal{Y}}$ for all $t \in (0, T]$ with some interval $\mathcal{Y} \subset \mathbb{R}$. In addition assume that the pricing functions $C^i(t, s, y)$ defined in (4.2) are continuously differentiable with respect to t and twice continuously differentiable with respect to s and y.

Clearly, these assumptions are satisfied by the usual stochastic volatility models like Heston's model or the Stein-Stein model. Furthermore it is well known that under Assumption 5.6 the pricing function $C^i : (0, T_i) \times (0, \infty) \times \mathcal{Y} \to \mathbb{R}$ of call C^i satisfies the parabolic differential equation

$$\frac{1}{2}s^2\sigma(y)^2\frac{\partial^2 C^i}{\partial s^2} + \rho s\sigma(y)\gamma(y)\frac{\partial^2 C^i}{\partial s \partial y} + \frac{1}{2}\gamma(y)^2\frac{\partial^2 C^i}{\partial y^2} + \\ + rs\frac{\partial C^i}{\partial s} + \beta(y)\frac{\partial C^i}{\partial y} + \frac{\partial C^i}{\partial t} = rC^i \qquad (5.10)$$

$$(t, s, y) \in (0, T_i) \times (0, \infty) \times \mathcal{Y}$$

with final condition $C^i(T_i, s, y) = (s - K_i)^+ \ \forall\, s \in (0, \infty)$ and appropriate boundary conditions. These findings will be used to show that the robust hedging problem (5.9) in fact has a solution.

Theorem 5.7. *Consider the optimization problem of finding a cost-optimal robust static super-replication strategy as stated in Definition 5.5. Assume that* $\Theta_1 = [0, T] \times \bar{\mathcal{Y}}$ *and* $\Theta_2 = [0, D]$. *If the financial market model satisfies Assumption 5.6, then a solution of the robust optimization problem (5.9) exists. Furthermore the set of solutions is convex and compact.*

Proof. By choosing a sufficiently large bond position, we observe that the feasible set of optimization problem (5.9) is nonempty. Hence, due to well-known theorems in linear semi-infinite optimization (see Theorem 1.12) it is sufficient to show that the objective function and the feasible set have no direction of recession in common. Assume that $d \in \mathbb{R}^{n+1}$ is such a direction of recession. In the following we will show that d must be equal to the zero vector which proves the theorem. By definition d satisfies

$$d_0 B_0 + \sum_{i=1}^{n} d_i C^i(0, S_0, Y_0, p_0) \leq 0$$

$$d_0 B_t + \sum_{C^i \in \mathcal{C}, T_i \geq t} d_i C^i(t, D, y, p) \geq 0 \quad \forall\, (t, y, p) \in \Theta_1 \times U \tag{5.11}$$

$$d_0 B_T + \sum_{C^i \in \mathcal{C}, T_i = T} d_i (s - K_i)^+ \geq 0 \quad \forall\, s \in \Theta_2.$$

In particular the second inequality holds for the parameter $p_0 \in U$. But then one can show in analogy to the proof of Theorem 5.2 that the set of constraints

$$d_0 B_t + \textstyle\sum_{C^i \in \mathcal{C}, T_i \geq t} d_i C^i(t, D, y, p_0) \geq 0 \quad \forall\, (t, y) \in \Theta_1 = [0, T] \times \bar{\mathcal{Y}}$$
$$d_0 B_T + \textstyle\sum_{C^i \in \mathcal{C}, T_i = T} d_i (s - K_i)^+ \geq 0 \quad \forall\, s \in \Theta_2 = [0, D]$$

implies the equalities

$$\begin{aligned} &\sum_{C^i \in \mathcal{C}, T_i \geq t} d_i C^i(t, s, y, p_0) = 0 \quad \forall\, (t, s, y) \in [0, T] \times \{D\} \times \bar{\mathcal{Y}} \\ &\sum_{C^i \in \mathcal{C}, T_i \geq t} d_i C^i(t, s, y, p_0) = 0 \quad \forall\, (t, s, y) \in \{T\} \times [0, D] \times \bar{\mathcal{Y}}. \end{aligned} \tag{5.12}$$

Hence the value of the static knock-out trading strategy d consisting of standard calls vanishes for all $(t, s, y) \in ([0, T] \times \{D\} \times \bar{\mathcal{Y}}) \cup (\{T\} \times [0, D] \times \bar{\mathcal{Y}})$. It is easy to see for arbitrary $(t, s, y) \in [0, T] \times [0, D] \times \bar{\mathcal{Y}}$, that $\sum_{C^i \in \mathcal{C}, T_i \geq t} d_i C^i(t, s, y, p_0)$ is the fair value of the static knock-out trading strategy d starting at time t given $S_t = s$ and $Y_t = y$. Due to (5.12) the payoff of this trading strategy is zero in both cases of a barrier hit and no barrier hit before maturity T. Thus, as the market model is arbitrage-free, the fair value of d at time t given $S_t = s$ and $Y_t = y$ must be equal to zero, which is the discounted expected future cash flow. This implies

$$\sum_{C^i \in \mathcal{C}, T_i \geq t} d_i C^i(t, s, y, p_0) = 0 \quad \forall\, (t, s, y) \in [0, T] \times [0, D] \times \bar{\mathcal{Y}}. \tag{5.13}$$

Now we group the standard calls $C^i \in \mathcal{C}$ with $K_i \geq D$ into sets $I_1, \ldots, I_r$ with equal maturities $\bar{T}_1 < \bar{T}_2 < \ldots < \bar{T}_r$. By the superposition principle, the function $c(t, s, y) := \sum_{i \in I_r} d_i C^i(t, s, y, p_0)$ satisfies the parabolic differential equation (5.10) with end condition

$$c(\bar{T}_r, s, y) = \sum_{i \in I_r} d_i (s - K_i)^+ \quad \forall s \in (0, \infty).$$

Further (5.13) implies that c vanishes on $(\bar{T}_{r-1}, \bar{T}_r) \times (0, D) \times \mathbb{O}$. Thus we can conclude by Mizohata's uniqueness theorem [71] that $c(t, s, y) = 0$ for all $(t, s, y) \in (\bar{T}_{r-1}, \bar{T}_r) \times (0, \infty) \times \mathbb{O}$. In particular this implies $\sum_{i \in I_r} d_i (s - K_i)^+ = 0$ for all $s \in (0, \infty)$ and hence $d_i = 0$ for $i \in I_r$. This argument is repeated recursively on the time strips $(\bar{T}_i, \bar{T}_{i+1}]$ such that all coefficients d_i have to vanish. □

The theorem shows that a solution of optimization problem (5.9) exists under mild conditions satisfied by stochastic volatility models commonly used in practical applications. In case a model does not satisfy Assumption 5.6, the existence can still be guaranteed by adding simple box constraints $\alpha_i^{\mathrm{lb}} \le \alpha_i \le \alpha_i^{\mathrm{ub}}$ to the description of optimization problem (5.9), because the minimization of a linear function on a closed convex and bounded set clearly has a solution.

However, in general an analytic derivation of the solution is not possible. Hence the next subsection will be devoted to the derivation of an algorithm for the numerical solution of the optimization problem.

5.2.2 Solving the Problem

To solve optimization problems (5.8) and (5.9) numerically, we first observe that these problems are linear semi-infinite optimization problems of the form

$$\begin{aligned} \min_{\alpha \in \mathbb{R}^{n+1}} \quad & c^{\mathrm{T}} \alpha \\ \text{s.t.} \quad & a_1(t, y, p)^{\mathrm{T}} \alpha \ge 0 \quad \forall\, (t, y, p) \in \Gamma_1 \subseteq \Theta_1 \times U \\ & a_2(s)^{\mathrm{T}} \alpha \ge b_2(s) \quad \forall\, s \in \Gamma_2 = [0, D] \\ & \alpha_i^{\mathrm{lb}} \le \alpha_i \le \alpha_i^{\mathrm{ub}}, \end{aligned} \tag{5.14}$$

where c, a_1, a_2 and b_2 denote suitable vectors and scalars, respectively. The additional box constraints allow to impose bounds on the hedge positions and guarantee the existence of a solution for arbitrary models not satisfying Assumption 5.6. Further note that the non-robust problem (5.8) can be obtained by setting $\Gamma_1 = \Theta_1 \times \{p_0\}$ and hence any algorithm for the solution of (5.14) can also be applied to solve the non-robust problem.

In general an algorithm solving problem (5.14) will replace the infinite number of constraints associated with the sets Γ_1 and Γ_2 by a discrete set of constraints. Let the discrete approximations of these index sets be denoted by $M_1 \subset \Gamma_1$ and $M_2 \subset \Gamma_2$, $|M_1|, |M_2| < \infty$. By neglecting the rest of the constraints, an optimal solution of the resulting discretized problem is in general not feasible for the original problem (5.14). To reduce this infeasibility one might employ a refinement of the meshes M_1, M_2 around nearly active constraints. This was the idea of Algorithm 2 which we applied to solve the robust static hedging problem in the Black–Scholes case. However, for typical stochastic volatility models the set Γ_1 is six-dimensional which prevents a local mesh refinement because of the curse of dimensionality. In these cases cutting plane discretizations are more suitable methods as for example presented in Goberna and Lopez [46]. Applying these methods to problem (5.14) leads to Algorithm 3.

Algorithm 3: Cutting Plane Discretization

Input: Let $M_1 \subset \Gamma_1$ and $M_2 \subset \Gamma_2$, $|M_1|, |M_2| < \infty$ be given initial grids and $(\epsilon_k)_{k\in\mathbb{N}}$ a sequence of non-negative numbers converging to zero. Further let $\mathrm{TOL} > 0$ be a suitable convergence tolerance and set $k = 0$.

Main Algorithm:

(S1) Calculate an optimal solution α^k of the discretized problem

$$\begin{aligned} \min_{\alpha\in\mathbb{R}^{n+1}} \quad & c^{\mathrm{T}}\alpha \\ \text{s.t.} \quad & a_1(t,y,p)^{\mathrm{T}}\alpha \geq 0 \quad \forall\,(t,y,p) \in \mathbf{M_1} \\ & a_2(s)^{\mathrm{T}}\alpha \geq b_2(s) \quad \forall\, s \in \mathbf{M_2} \\ & \alpha_i^{\mathrm{lb}} \leq \alpha_i \leq \alpha_i^{\mathrm{ub}} \end{aligned} \tag{5.15}$$

(S2) Determine the constraint violation (slack) of α^k for problem (5.14) by minimizing the slack–functions at α^k:

$$\delta_1 = \min_{(t,y,p)\in\Gamma_1} a_1(t,y,p)^{\mathrm{T}}\alpha^k, \qquad \delta_2 = \min_{s\in\Gamma_2} a_2(s)^{\mathrm{T}}\alpha^k - b_2(s) \tag{5.16}$$

In the process of these minimizations identify nonempty finite sets Ψ_1, Ψ_2 of ϵ_k-minimizers satisfying

$$\begin{aligned} a_1(t,y,p)^{\mathrm{T}}\alpha^k &\leq \delta_1 + \epsilon_k \quad \forall\,(t,y,p) \in \Psi_1 \\ a_2(s)^{\mathrm{T}}\alpha^k - b_2(s) &\leq \delta_2 + \epsilon_k \quad \forall\, s \in \Psi_2. \end{aligned}$$

If $\min\{\delta_1, \delta_2\} \geq -\mathrm{TOL}$ then STOP.

(S3) Add the ϵ_k-minimizers of the slack functions (the most violating constraints) to M_1, M_2 by setting $M_1 \leftarrow M_1 \cup \Psi_1$, $M_2 \leftarrow M_2 \cup \Psi_2$. Further set $k \leftarrow k+1$ and go to step (S1).

To solve the robust optimization problem, Algorithm 3 successively solves a sequence of linear optimization problems (S1) and nonlinear optimization problems (S2). The solutions of the linear optimization problems can easily be computed with available linear programming solvers. In contrast the solution of the global optimization problems in step (S2) is more challenging. For most models it is easy to see that these minimization problems are smooth except some lines of non-differentiability in time-direction occuring at the maturities T_i of the calls included in the hedge portfolio (a more detailed discussion can be found for the Black–Scholes case at the end of

Subsection 5.1.2). As these lines of non-differentiability are well known in advance, the minimization can be carried out separately on each smooth subregion. To compute the respective global minima one can for example successively apply nonlinear optimization algorithms with changing (random) start iterate. The outcome of these minimizations may be several ϵ_k-optimal solutions which can be gathered in the sets Ψ_1, Ψ_2. Appendix B illustrates in the form of Matlab code how the method described above can be implemented in practice.

From a financial point of view, the algorithm first calculates a hedge portfolio in step (S1) guaranteeing the super-replication property for the cases of a barrier hit $(t, y, p) \in M_1$ and the possible stock prices $s \in M_2$ in case of no barrier hit. In step (S2) the algorithm computes the worst case hedge error of the portfolio α^k for all possible states of the market not considered by the sets M_1 and M_2. The most violating states are then added to the sets M_1 and M_2 leading to a more robust solution in the next iteration. This procedure is repeated recursively until the portfolio delivers a worst case hedge error smaller than TOL. The next theorem shows, that this iterative procedure converges to the desired robust hedge portfolio under suitable assumptions.

Theorem 5.8. *Assume $\exists\, M > 0$ such that $\|a_1(t, y, p)\| \leq M \;\forall\, (t, y, p) \in \Gamma_1$. If the feasible set of problem (5.14) is nonempty and $\min_{(t,y,p)\in\Gamma_1} a_1(t, y, p)^{\mathrm{T}}\alpha^k$ exists at each iteration of Algorithm 3, then every limit point of the sequence $(\alpha^k)_k$ is an optimal solution of problem (5.14).*

Proof. Since the feasible set of problem (5.14) is nonempty, closed, convex and due to the box constraints compact, a solution of problem (5.14) exists. By the same argument, an optimal solution of the discretized problem (5.15) exists for arbitrary sets M_1 and M_2.

It is easy to see that $s \mapsto a_2(s)$ is a continuous function such that $\|a_2(s)\|$ is bounded for $s \in [0, D]$. The continuity also implies that $\min_{s\in[0,D]} a_2(s)^{\mathrm{T}}\alpha^k - b_2(s)$ exists for arbitrary iterates α^k. Hence Algorithm 3 is well defined. Applying the general convergence theory of linear semi-infinite optimization (see e.g. Goberna and Lopez [46], Theorem 11.2), the theorem immediately follows. □

Note that the boundedness assumption of Theorem 5.8 as well as the existence of the minimum are trivially fulfilled if $\Gamma_1 \subseteq \Theta_1 \times P$ is a compact set. However, as illustrated in Example 4.5, the set Θ_1 is closed but for some models not bounded. Hence $\Theta_1 \times P$ is not necessarily bounded as well. In these cases, from a numerical point of view, it is necessary to restrict $\Theta_1 \times P$ to a compact set Γ_1 in order to carry out the minimizations in step (S2) of the algorithm. This restriction can also be justified from a financial perspective: An unboundedness of Θ_1 can only occur in the volatility direction y of Θ_1 leading to extreme volatility states like $+\infty$ which is of no practical interest. Thus it is natural to exclude extreme volatility states, for example one might choose in case of Heston's model $\Gamma_1 = [0, T] \times [0, y_{\max}] \times P$ instead of $[0, T] \times [0, \infty) \times P$ with an appropriate upper bound $y_{\max} > 0$.

Further note that the feasible set of problem (5.14) is nonempty if the box constraints or hedge instruments are chosen appropriately. For example the condition $\alpha_0^{\text{lb}} \le (D - K)^+/B_T \le \alpha_0^{\text{ub}}$ and $\alpha_i^{\text{lb}} \le 0 \le \alpha_i^{\text{ub}}$, $i = 1, \dots, n$, assures the existence of a static super-replication portfolio, because then the strategy $\alpha_0 = (D - K)^+/B_T$, $\alpha_1 = \dots = \alpha_n = 0$ is feasible. But also the simple hedging strategy of buying a standard call with the same strike and maturity as the barrier option is a feasible strategy if the box constraints do not exclude it. Hence the assumptions of Theorem 5.8 are fulfilled for all cases of practical interest such that a solution of problem (5.14) exists and any limit point of the sequence $(\alpha^k)_k$ is an optimal hedge portfolio.

5.2.3 A Detailed Example

Although the previous findings are satisfactory from a theoretical point of view, the question remains how the robustified portfolio performs in practice. In this subsection we investigate this question with a detailed numerical example. But before considering the robust problem, we first illustrate the performance of Algorithm 3 by solving the non-robust static hedging problem which was also the basis of the computations in Subsection 3.3.2. Afterwards, we analyze the sensitivity of the non-robust portfolio with respect to changes in the model parameters. The analysis shows that the skew risk cannot be neglected and may lead to losses which amount to multiples of the fair value of the barrier option. Finally we successively robustify the static hedge portfolio against changes of the volatility surface and quantify the cost of this robustification.

To make our results comparable with those of Subsection 3.3.2, we specify the general stochastic volatility model (4.1) as the well-known Heston model (see Example 4.5). Besides a good fit of the volatility surface this model has the particular advantage that a closed form solution for the price of standard calls is readily available (see Section 3.2 or Heston [50]) to significantly speed up computations.

The Heston model parameters are chosen as follows: The risk-free interest rate is defined to be $r = 5.5\%$, the dividend yield $\delta = 2.5\%$, the start variance $Y_0 = 0.04$, the long run mean of the variance $\theta = 0.04$, the mean reversion speed $\kappa = 1.5$, the volatility of volatility $\xi = 0.2$ and the correlation $\rho = -0.5$.

The goal is to hedge an up-and-out call with strike $K = 2750$, barrier $D = 3300$ and a maturity of $T = 1$ year by a portfolio consisting of the calls listed in Table 5.8. The value of the underlying at time $t = 0$ is assumed to be $S_0 = 2750$. For this set of parameters, the fair value of the up-and-out call is 1.60% of the underlying S_0 (see Table 3.5).

	C^1	C^2	C^3	C^4	C^5	C^6	C^7	C^8	C^9	C^{10}
T_i	1.00	1.00	1.00	1.00	1.00	0.75	0.75	0.75	0.50	0.50
K_i	2750	3300	3350	3450	3600	3300	3400	3600	3300	3500

Table 5.8: Standard calls C^i included in the hedge portfolio

As mentioned in the Black–Scholes case, the set of calls $\{C^1, \ldots, C^{10}\}$ satisfies Assumption 4.2. Hence, by Theorem 4.4, the stochastic optimization problem (4.3) is equivalent to the semi-infinite optimization problem (5.8) with $\Theta_1 = [0, T] \times [0, \infty)$ and $\Theta_2 = [0, D]$ (see Example 4.5). If we define the parameter vector $p = (\kappa, \theta, \xi, \rho)^{\mathrm{T}}$, it is easy to see that Heston's model satisfies the assumptions of Definition 5.5 as well as Assumption 5.6. Accordingly, Theorem 5.7 implies that the robust optimization problem (5.9) has a solution and that the set of optimal solutions is convex and compact.

To ensure that the discretized subproblems (5.15) also have a solution in each iteration of Algorithm 3 we impose the simple bounds $\alpha_i^{\mathrm{lb}} = -50$ and $\alpha_i^{\mathrm{ub}} = 50$ on the portfolio positions. These bounds further guarantee that the trading strategy $\alpha_1 = 1$, $\alpha_i = 0, i \neq 1$, solely consisting of the call C^1 with the same strike and maturity as the barrier option is feasible for problem (5.14). Thus Theorem 5.8 implies for compact sets $\Gamma_1 \subseteq \Theta_1 \times U$ that every limit point of the sequence generated by Algorithm 3 is an optimal solution of optimization problem (5.14).

The Non-Robust Static Hedging Strategy The non-robust optimization problem is a special case of problem (5.14) by setting $U = \{p_0\}$ with $p_0 = (\kappa_0, \theta_0, \xi_0, \rho_0) := (1.5, 0.04, 0.2, -0.5)^{\mathrm{T}}$. In order to solve the resulting problem numerically, we have to restrict the unbounded set $\Theta_1 = [0, T] \times [0, \infty)$ to a suitable compact subset. From a financial point of view it is reasonable to bound the variance in case of a barrier hit by 100% such that we replace $\Theta_1 \times \{p_0\}$ by $\Gamma_1 := [0, T] \times [0, 1] \times \{p_0\}$.

In terms of Algorithm 3 we choose a first discretization M_1, M_2 of Γ_1, Γ_2 consisting of a total of 252 grid points. For this setup Table 5.9 shows that the algorithm terminates after 13 iterations with a worst case hedge error $\min\{\delta_1, \delta_2\}$ satisfying the prespecified tolerance TOL $= 10^{-7}$. Here the cost $\Pi_0(\alpha^k)$ as well as the worst case hedge error (slack) of the portfolio α^k are listed in percent of the underlying S_0. Obviously the cost of the hedge portfolio converges to approximately 1.80% which is not much more expensive than the fair value of the barrier option (1.60%). In particular the price difference is less than half of the typical bid-ask spread for barrier options in the OTC market.

The optimal solution of the problem is presented in Table 5.10. The optimal hedge portfolio is very close to the one computed by solving the stochastic optimization problem (4.3) with the Monte Carlo-based Algorithm 1 (see Table 3.4). Again, this is not surprising since the semi-infinite problem representation is equivalent to the original stochastic super-replication problem. However, due to the semi-infinite equivalence and by exploiting this structure in Algorithm 3, we are now able to compute the hedge portfolio with an even higher accuracy in a few seconds instead of several minutes or even hours.

Figure 5.6 (a) gives some further insight into the structure of the hedge error in case of a barrier hit. It is clearly visible that the optimal strategy is a super-replication strategy guaranteeing a payoff greater or equal to zero on the barrier (neglecting the

Iteration	Slack	$\Pi_0(\alpha^k)$	$\lvert M_1\rvert + \lvert M_2\rvert$	ϵ_k
0	−2.037945e−001	1.419938e−002	252	1.00000e−002
1	−6.515309e−003	1.768397e−002	266	1.00000e−003
2	−1.378396e−003	1.791617e−002	273	1.00000e−004
3	−2.007605e−003	1.792881e−002	275	1.00000e−005
4	−7.201896e−004	1.796985e−002	285	1.00000e−006
5	−4.653767e−004	1.800102e−002	288	1.00000e−007
6	−9.036166e−005	1.800403e−002	298	1.00000e−008
7	−2.090027e−004	1.800496e−002	302	1.00000e−008
8	−1.616535e−005	1.800862e−002	304	1.00000e−008
9	−4.707759e−005	1.800986e−002	311	1.00000e−008
10	−1.125056e−005	1.801157e−002	313	1.00000e−008
11	−3.552461e−006	1.801166e−002	322	1.00000e−008
12	−1.999220e−006	1.801171e−002	329	1.00000e−008
13	−3.458727e−008	1.801171e−002	334	1.00000e−008

Table 5.9: Iteration process for the non-robust problem

	C^1	C^2	C^3	C^4	C^5	C^6	C^7	C^8	C^9	C^{10}
T_i	1.00	1.00	1.00	1.00	1.00	0.75	0.75	0.75	0.50	0.50
K_i	2750	3300	3350	3450	3600	3300	3400	3600	3300	3500
α_i	1.00	−25.29	29.28	−5.97	1.42	0.35	−0.26	−0.42	0.23	−0.23

Table 5.10: Optimal portfolio weights α_i for the non-robust problem

minimal hedge error of -3.4×10^{-8} listed in Table 5.9). Shortly before the maturity $T = 1$ of the barrier option the value of the hedge portfolio bends upwards to $(D - K)/S_0 = 20\%$ to super-replicate the barrier option in case of no barrier hit. Furthermore it is easy to recognize the lines of non-differentiability along the maturities of the standard calls in the portfolio ($t = 0.5$ and $t = 0.75$). In summary, the optimal hedge portfolio presented in Table 5.10 behaves as expected by offering protection for a wide range of volatility states and hitting times.

However, this well-behaved evolvement of the hedge error completely changes if the model parameters p in case of a barrier hit differ from those used for the computation of the hedge portfolio ($p_0 = (1.5, 0.04, 0.2, -0.5)^{\mathrm{T}}$). Figure 5.6 (b) illustrates graphically that the value of the hedge portfolio and hence the hedge error on the barrier is extremely sensitive to changes in the model parameters. The perturbation from p_0 to $p = (1.4, 0.025, 0.25, -0.6)^{\mathrm{T}}$ leads to possibly huge hedging losses of up to 3.0% which is clearly unacceptable in comparison to the price of the barrier option (1.60%). This model parameter uncertainty is caused by the strong nonlinearity of the call prices and hence the function $a_1(t, y, p)$ in problem (5.14).

To better quantify the effect of the model parameter uncertainty, we analyze the worst case hedging error δ_1 (see (5.16) in Algorithm 3) of the non-robust hedge port-

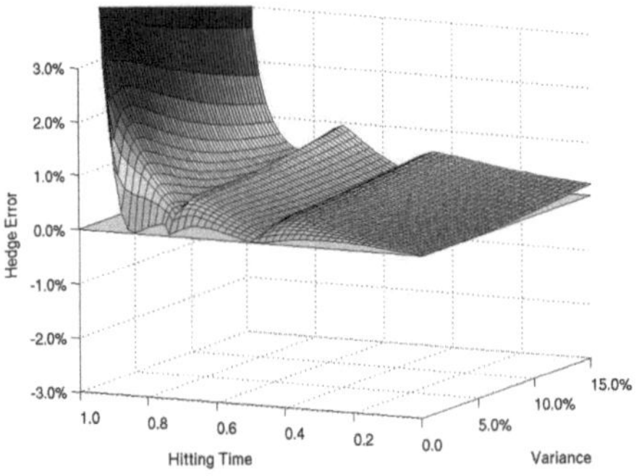

(a) Parameter $p_0 = (1.5, 0.04, 0.2, -0.5)^{\mathrm{T}}$

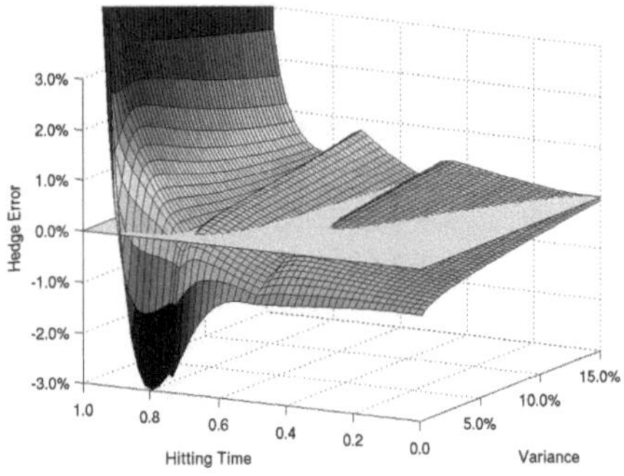

(b) Perturbed $p = (1.4, 0.025, 0.25, -0.6)^{\mathrm{T}}$

Figure 5.6: Hedge error on the barrier for the non-robust portfolio

folio for varying model parameter sets U. Starting with the non-robust case $U = \{p_0\}$ we successively increase U to capture the risk of more wildly moving model parameters. For this purpose it is sufficient to model U as multi-dimensional intervals $[\kappa_{\min}, \kappa_{\max}] \times [\theta_{\min}, \theta_{\max}] \times [\xi_{\min}, \xi_{\max}] \times [\rho_{\min}, \rho_{\max}]$ with uniformly increasing size $\Delta := \kappa_{\max} - \kappa_{\min} = \sqrt{\theta_{\max}} - \sqrt{\theta_{\min}} = \xi_{\max} - \xi_{\min} = \rho_{\max} - \rho_{\min}$ of the individual one-dimensional parameter sets. Of course, in addition the points in U must satisfy the Heston box constraints $\kappa \geq 0$, $\theta \geq 0$, $\xi \geq 0$, $-1 \leq \rho \leq 1$ as well as the cone constraint $\kappa\theta - \xi^2/2 \geq 0$ to guarantee the positivity of the variance process $(Y_t)_t$. For example, the multi-dimensional interval with size $\Delta = 20\%$ around $p_0 = (1.5, 0.04, 0.2, -0.5)^{\mathrm{T}}$ is given by $[1.4, 1.6] \times [0.01, 0.09] \times [0.1, 0.3] \times [-0.6, -0.4]$. The set U we consider is the intersection of this interval with the box and cone constraints. Note that we do not need to explicitely construct U, but instead we can implicitely include its definition as constraints in the nonlinear optimization problem (5.16). The resulting worst case hedge errors for a variety of sizes Δ of the uncertainty set are shown in Table 5.11.

Size Δ	0%	5%	10%	15%	20%	30%	40%
WC-Error	0.0%	−1.20%	−2.90%	−3.90%	−4.83%	−5.82%	−6.91%

Table 5.11: Worst case hedge error δ_1 of the non-robust portfolio for model parameter sets U around p_0 with varying size Δ

The numbers shown in Table 5.11 precisely quantify the model parameter uncertainty of the non-robust hedge portfolio. For a fixed model parameter p_0 ($\Delta = 0\%$) the portfolio offers perfect protection over the lifetime of the barrier option. However, if model parameters only change slightly the nonlinearity of the call option prices leads to hedging losses that can be multiples of the fair value of the barrier option. For example the perturbed parameter p in Figure 5.6 (b) is contained in the uncertainty set U with size $\Delta = 20\%$, but the graphically visible worst case loss of approximately -3.0% is not even the largest loss which can occur within the whole set U (WC-Error=-4.83%). As implied model parameters p change daily, the model parameter uncertainty (skew-risk) hence cannot be neglected and must be included in the design of static hedge portfolios. This will be the focus for the remainder of this section.

Adding Robustness to the Hedge Portfolio Due to the strong effect of model parameter uncertainty we now aim at adding robustness to the static hedge portfolio. By definition the robust optimization problems (5.9) and (5.14) allow to easily take model parameter uncertainty into account by means of a set U in which the parameters p are allowed to vary without losing the super-replication property.

Hence the question arises of how to choose a suitable set U in practice. Of course the set should be chosen such that the implied model parameters p stay within U over the lifetime of the barrier option. At first sight this does not seem like an easy task, because it requires to predict the future. But a quick analysis of historical implied model parameters shows that the movement of the parameters is not arbitrary. This can for example be observed in Figure 5.7, which shows the calibrated model parameters for the case of the EURO STOXX 50 index over a time span of 16 months. Although the parameters $p = (\kappa, \theta, \xi, \rho)^T$ vary stochastically over time, the plot reveals some inherent structure. In particular the implied parameters stay within certain bounds which allows us to capture them in appropriate robustness intervals. This also empirically justifies the choice of U as the intersection of the multidimensional interval $[\kappa_{\min}, \kappa_{\max}] \times [\theta_{\min}, \theta_{\max}] \times [\xi_{\min}, \xi_{\max}] \times [\rho_{\min}, \rho_{\max}]$ with the set of points satisfying the Heston constraints.

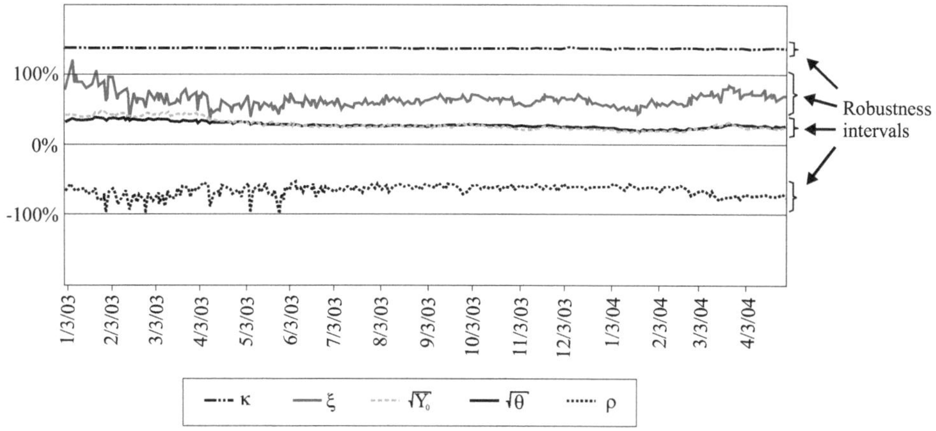

Figure 5.7: Implied Heston parameters over time for the EURO STOXX 50 index resulting from daily calibrations with soft penalty. Source: Hans Buehler [20], Deutsche Bank AG

For our computations we choose a size of $\Delta = 10\%$ for the intervals describing the uncertainty set U, that means the super-replication property is preserved if the future Heston parameters κ, θ, ξ and ρ vary within a ten percent band centered around the initial model parameters p_0. As in the non-robust case, the initial variance and hence $\sqrt{Y_0}$ is allowed to float within the interval $[0, 100\%]$ leading to the set $\Gamma_1 = [0, T] \times [0, 1] \times U$. We start Algorithm 3 with a first discretization M_1, M_2 of the sets Γ_1, Γ_2 consisting of a total of 6264 grid points. The resulting iteration process is shown in Table 5.12.

Iteration	Slack	$\Pi_0(\alpha^k)$	$\|M_1\| + \|M_2\|$	ϵ_k
0	−3.966997e−001	1.116424e−002	6264	1.00000e−002
1	−2.880382e−001	1.550770e−002	6271	1.00000e−003
2	−1.183795e−002	2.003119e−002	6284	1.00000e−004
3	−5.024301e−003	2.009495e−002	6288	1.00000e−005
4	−2.684907e−003	2.016812e−002	6297	1.00000e−006
5	−1.545740e−003	2.022707e−002	6313	1.00000e−007
6	−8.765325e−004	2.023450e−002	6320	1.00000e−008
7	−8.886678e−004	2.026973e−002	6328	1.00000e−008
8	−1.681279e−004	2.027408e−002	6335	1.00000e−008
9	−1.523032e−004	2.028087e−002	6339	1.00000e−008
10	−4.985848e−005	2.028104e−002	6347	1.00000e−008
11	−4.441402e−005	2.028107e−002	6356	1.00000e−008
12	−3.175284e−005	2.028238e−002	6370	1.00000e−008
13	−9.252064e−005	2.028338e−002	6378	1.00000e−008
14	−1.244341e−006	2.028352e−002	6381	1.00000e−008
15	−2.207395e−005	2.028361e−002	6392	1.00000e−008
16	−8.364235e−006	2.028369e−002	6395	1.00000e−008
17	−3.260814e−006	2.028393e−002	6398	1.00000e−008
18	−2.688489e−007	2.028397e−002	6404	1.00000e−008

Table 5.12: Iteration process for the robust problem, $\Delta = 10\%$

The algorithm terminates after 18 iterations with the optimal solution presented in Table 5.13 and a worst case hedging loss less than $\text{TOL} = 10^{-6}$. The cost of the portfolio converges to 2.03% which is still cheap for a static super-replication portfolio with a 10%-robustness against model parameter uncertainty. In particular the computed portfolio, which is 23 basis points more expensive than the non-robust portfolio (1.80%), completely eliminates the risk of suffering the worst case loss of −2.90% (see Table 5.11). Compared to this loss the additional premium seems rather low.

Of course the robustness of the hedge portfolio listed in Table 5.13 is limited to model parameter changes within the prespecified intervals with size 10% around p_0. However, traders might prefer more conservative hedge portfolios offering protection for an even wider range of model parameters. For this purpose, Table 5.14 illustrates

	C^1	C^2	C^3	C^4	C^5	C^6	C^7	C^8	C^9	C^{10}
T_i	1.00	1.00	1.00	1.00	1.00	0.75	0.75	0.75	0.50	0.50
K_i	2750	3300	3350	3450	3600	3300	3400	3600	3300	3500
α_i	1.00	−22.02	25.46	−6.65	2.71	−0.04	0.60	−0.94	0.37	−0.43

Table 5.13: Optimal portfolio weights α_i for the robust problem, $\Delta = 10\%$

the cost of static hedge portfolios with varying degree of robustness. As expected the cost increases with an increasing size of the robustness intervals. But still the cost is surprisingly low in comparison to the potential hedging losses if the model parameter uncertainty is not taken into account (see Table 5.11). Further note that as in the Black–Scholes case the cost increases linearly in the size Δ of the uncertainty set.

Size Δ	0%	5%	10%	20%	30%	40%
$\sqrt{Y_{\min}}$	0.0%	0.0%	0.0%	0.0%	0.0%	0.0%
$\sqrt{Y_{\max}}$	100.0%	100.0%	100.0%	100.0%	100.0%	100.0%
$\sqrt{\theta_{\min}}, \xi_{\min}$	20.0%	17.5%	15.0%	10.0%	10.0%	10.0%
$\sqrt{\theta_{\max}}, \xi_{\max}$	20.0%	22.5%	25.0%	30.0%	40.0%	50.0%
$\kappa_{\min}$	150.0%	147.5%	145.0%	140.0%	130.0%	120.0%
$\kappa_{\max}$	150.0%	152.5%	155.0%	160.0%	160.0%	160.0%
$\rho_{\min}$	−50.0%	−52.5%	−55.0%	−60.0%	−70.0%	−80.0%
$\rho_{\max}$	−50.0%	−47.5%	−45.0%	−40.0%	−40.0%	−40.0%
Cost	**1.80%**	**1.91%**	**2.03%**	**2.14%**	**2.25%**	**2.36%**

Table 5.14: Cost of optimal hedge portfolios with varying degree of robustness

So far we treated the short term volatility parameter $\sqrt{Y_0}$ separately from the other Heston parameters κ, θ, ξ and ρ. This is also visible in Table 5.14 where we can observe that $\sqrt{Y_0}$ is allowed to vary in the robustness interval [0, 100%] while all other parameters are bound to the uncertainty set with size Δ. This separate treatment stems from the super-replication property in the stochastic volatility model which theoretically requires the static hedge portfolio to provide a payoff greater or equal to the barrier option in all possible volatility states. But while the Heston model assumes that the barrier can be hit at any non-negative volatility state, Figure 5.7 reveals that in practice the implied short term volatility $\sqrt{Y_0}$ can be captured within prespecified bounds, too. This is also clear if we think of the volatility surfaces generated by the Heston model. A short term volatility of up to 100% would lead to a surface with implied volatilities of nearly 100% for options close to maturity and much lower implied volatilities afterwards approaching the long term volatility level $\sqrt{\theta}$. Clearly, such a market scenario is unrealistic and can be excluded from the analysis.

By excluding market scenarios we further relax the theoretical super-replication property $\Pi_T(\alpha) \geq C_{\text{uo}}$ (a.s.) for the Heston model with fixed model parameter p_0 (see the equivalence Theorem 4.4), but focus on what is really important for the design of a good robust static hedge portfolio: The value of the hedge portfolio must be greater or equal to zero in case of a barrier hit. As the value of the hedge portfolio at the time of a barrier hit is given by the future call prices (which are part of the future volatility surface), it is sufficient to require the super-replication property to hold for a set of possible future implied model parameters. And this property is not reflected by an almost-surely constraint but precisely by the semi-infinite constraints in problem (5.14) with an interval-type set $\Gamma_1 := [0, T] \times [Y_{\min}, Y_{\max}] \times U \subsetneq \Theta_1 \times U$, where the short-term volatility is also constrained by appropriate lower and upper bounds $\sqrt{Y_{\min}}$, $\sqrt{Y_{\max}}$. Note that the arguments above are consistent with the interpretation of the dual problem derived in Section 4.2. In this problem the fit of the market prices turned out to be decisive for the measures under consideration. In a similar spirit, we now "abuse" the Heston model to generate likely market prices (volatility surface scenarios) and are not interested in any other property of the model.

If we relax the extreme robustness of the hedge portfolio against changes of the short term volatility parameter to a more practical uncertainty interval, we reduce the number of constraints in optimization problem (5.9). Hence the resulting optimal hedge portfolios should be cheaper than those listed in Table 5.14. To precisely quantify the reduction of the portfolio cost we restart Algorithm 3 with the new uncertainty set $\Gamma_1 := [0, T] \times [Y_{\min}, Y_{\max}] \times U \subsetneq \Theta_1 \times U$ with varying size $\Delta = \sqrt{Y_{\max}} - \sqrt{Y_{\min}} = \kappa_{\max} - \kappa_{\min} = \sqrt{\theta_{\max}} - \sqrt{\theta_{\min}} = \xi_{\max} - \xi_{\min} = \rho_{\max} - \rho_{\min}$. The results are presented in Table 5.15.

Size Δ	5%	10%	15%	20%	30%	Non-unif.
$\sqrt{Y_{\min}} = \sqrt{\theta_{\min}}$	17.5%	15.0%	12.5%	10.0%	10.0%	17.5%
$\sqrt{Y_{\max}} = \sqrt{\theta_{\max}}$	22.5%	25.0%	27.5%	30.0%	40.0%	42.5%
$\xi_{\min}$	17.5%	15.0%	12.5%	10.0%	10.0%	45.0%
$\xi_{\max}$	22.5%	25.0%	27.5%	30.0%	40.0%	105.0%
$\kappa_{\min}$	147.5%	145.0%	142.5%	140.0%	130.0%	130.0%
$\kappa_{\max}$	152.5%	155.0%	157.5%	160.0%	160.0%	150.0%
$\rho_{\min}$	−52.5%	−55.0%	−57.5%	−60.0%	−70.0%	−95.0%
$\rho_{\max}$	−47.5%	−45.0%	−42.5%	−40.0%	−40.0%	−55.0%
Cost	**1.62%**	**1.73%**	**1.84%**	**1.95%**	**2.07%**	**2.08%**

Table 5.15: Cost of robust portfolios with restricted volatility robustness

Obviously, the reduced volatility robustness leads to significantly cheaper hedge portfolios. In particular the price difference of a 20% robust portfolio and the fair value of the barrier option (1.60%) is still within the typical bid-ask spread for barrier options in the OTC market. As an example Table 5.16 shows the hedge portfolio for the uncertainty set with diameter $\Delta = 15\%$. The cost of this portfolio amounts to 1.84%, which is just slightly more expensive than the cost 1.80% of the non-robust portfolio (see Table 5.10) only offering protection against changes of the short-term volatility parameter.

Finally note that the last column of Table 5.15 displays the cost of the (very conservative) robust hedge portfolio corresponding to the robustness intervals with non-uniform diameter displayed in Figure 5.7. Even though the uncertainty set covers most of the implied model parameters during a one year time frame, the cost of 2.08% is still surprisingly low. The surcharge $2.08\% - 1.60\% = 0.48\%$ on the fair value (1.60%) of the barrier option can be regarded as a risk premium for the model parameter uncertainty, and is thus also of practical relevance for dynamic trading strategies.

	C^1	C^2	C^3	C^4	C^5	C^6	C^7	C^8	C^9	C^{10}
T_i	1.00	1.00	1.00	1.00	1.00	0.75	0.75	0.75	0.50	0.50
K_i	2750	3300	3350	3450	3600	3300	3400	3600	3300	3500
α_i	1.00	−30.42	37.95	−9.13	0.47	0.04	0.31	−0.29	0.01	0.10

Table 5.16: Optimal portfolio weights α_i for the robust problem with relaxed short term volatility robustness, $\Delta = 15\%$

Before we move on to extensions of the robust static hedging approach in the next chapters, we briefly summarize the obtained results. The analysis of model parameter uncertainty in the Black–Scholes model as well as Heston's stochastic volatility model has shown that this kind of uncertainty can lead to extreme hedging losses. Due to the strong nonlinearity of the call option prices even small perturbations of the model parameters can lead to huge losses amounting to multiples of the fair value of the barrier option. But while this sensitivity of static hedge portfolios has been observed by various authors in simulation-based studies, we presented a method to exactly quantify the uncertainty by solving a nonlinear optimization problem.

Furthermore, the semi-infinite representation of the static super-replication approach allowed us for the first time to incorporate robustness against volatility and skew risk into the design of the static hedging strategy. The degree of robustness can simply be steered by the size Δ of the uncertainty set containing the expected future implied model parameters. As traders usually have a very good feeling of how implied model parameters change over time, an appropriate parameter Δ can easily be chosen.

After proving the existence of a robust static hedging strategy, we presented an algorithm to numerically solve the optimization problem. By successively solving a sequence of linear and nonlinear optimization problems, the algorithm generates iterates converging to the optimal static hedge portfolio in the continuous time financial market model. Based on the semi-infinite problem structure, the computation time for the optimal static hedging strategy was reduced drastically in comparison to the Monte Carlo-based approach. Detailed numerical examples showed the applicability of our framework. Although the model parameter uncertainty of static hedge portfolios for barrier options is huge, it can be eliminated by surprisingly low cost. In particular our method allows to compute robust static hedge portfolios with a price difference to the fair value of barrier options that is within the typical bid-ask-spread observed in the OTC market.

By gaining these insights it is also possible to identify what makes up a good model for the proposed static hedging approach. First of all, it is important that the model calibrates well to volatility surfaces, because the fit implies how close the price of the hedge portfolio at the time of a barrier hit is to the real price in the market. Furthermore, it is desired to choose the parameter uncertainty sets as small as possible, because larger uncertainty sets imply more constraints and hence more expensive hedge portfolios. Thus the implied model parameters should not vary too wildly over time. This means that a model is not a good model for the static super-replication approach if it has thousands of parameters, but if it leads to a good fit with stable parameters over time.

Heston's stochastic volatility model combines both desirable properties such that it is well suited for the computation of static hedge portfolios. Based on the optimization framework it is possible to identify the best super-replication portfolio consisting of calls that are really traded in the market. The numerical results showed, that in contrast to other static hedging approaches the upper bounds we obtain are very sharp. These sharp bounds stem from the fact that the presented approach exploits the relation of call prices for various strikes and maturities (the volatility surface) as much as possible. In contrast to this, completely model-independent static hedging approaches neglect the available information and consequently lead to very conservative bounds which are not competitive if compared to the ones presented in this chapter.

6 Modifications and Extensions

The intuitiveness of the static super-replication approach lends itself to several modifications and extensions which are presented in this chapter. First of all, Section 6.1 analyzes the concept of robust static sub-replication in the Black–Scholes and Heston's stochastic volatility model. With a slight modification of the algorithms developed so far, this approach allows to compute robust lower bounds for the price of barrier options. Combined with the super-replication price derived in Chapter 5 this leads to robust static bounds which are surprisingly tight regarding the incorporated robustness against changes of the volatility surface.

In a second step Section 6.2 addresses the question of jump risk. To avoid potential losses caused by jumps of the underlying we additionally robustify the hedge portfolio against stock prices above the barrier. Although this increases the dimension of the constraint system, we can prove that the problem complexity can be reduced significantly by applying the maximum principle for parabolic partial differential equations. The reduced problem can be interpreted as *moving the barrier* which demonstrates the practicability of the presented approach. Numerical results underline that jump risk can be controlled by an appropriate choice of the hedge positions.

Finally, Section 6.3 illustrates that the findings derived throughout this book are not limited to the case of statically hedging an up-and-out call. Instead, the semi-infinite problem formulation can be generalized to various other barrier options, including knock-in calls and puts, options with non-zero rebates and non-constant or even discrete barriers.

6.1 Sub-Replication and Robust Static Bounds

The results presented so far focused on the super-replication problem (3.1) which leads to upper bounds for the price of the target barrier option. However, it is well known in the literature of financial mathematics (see Section 1.2) that also lower bounds can be derived by considering the analogous sub-replication problem

$$\max_{\phi \in \mathcal{S}_{\mathrm{KO}}} \Pi_0(\phi) \quad \text{s.t.} \quad \Pi_T(\phi) \leq C_{\mathrm{uo}} \text{ (a.s.).} \tag{6.1}$$

Instead of minimizing the portfolio cost over the set of strategies super-replicating the barrier option, we now take the buyer's perspective and try to identify the most expensive portfolio providing a value which is always less than or equal to the payoff of the up-and-out call. For the general stochastic volatility model described in Assumption

4.1 this leads to the optimization problem

$$\max_{\alpha \in \mathbb{R}^{n+1}} \alpha_0 B_0 + \sum_{i=1}^{n} \alpha_i C^i(0, S_0, Y_0)$$

$$\text{s.t.} \quad \Pi_T(\alpha) = \alpha_0 B_T + \frac{B_T}{B_{\tau \wedge T}} \sum_{i=1}^{n} \alpha_i C^i(\tau \wedge T, S_{\tau \wedge T}, Y_{\tau \wedge T}) \leq C_{\text{uo}} \text{ (a.s.)} \tag{6.2}$$

$$\begin{aligned} dS_t &= (r - \delta) S_t dt + \sigma(Y_t) S_t d W_t^1 \\ dY_t &= \beta(Y_t) dt + \gamma(Y_t) d W_t^2 \\ dB_t &= r B_t dt. \end{aligned}$$

In analogy to the derivation of Theorem 4.4, problem (6.2) can be proven to be equivalent to the linear semi-infinite optimization problem

$$\max_{\alpha \in \mathbb{R}^{n+1}} \alpha_0 B_0 + \sum_{i=1}^{n} \alpha_i C^i(0, S_0, Y_0)$$

$$\begin{aligned} \text{s.t.} \qquad \alpha_0 B_t + \sum_{C^i \in \mathcal{C}, T_i \geq t} \alpha_i C^i(t, D, y) &\leq 0 \quad \forall\, (t, y) \in \Theta_1 \\ \alpha_0 B_T + \sum_{C^i \in \mathcal{C}, T_i = T} \alpha_i (s - K_i)^+ &\leq (s - K)^+ \quad \forall\, s \in \Theta_2, \end{aligned} \tag{6.3}$$

where Θ_1 and Θ_2 are defined as in the super-replication case. Furthermore, the existence results derived in Chapter 5 also carry over to the sub-replication setting. To numerically compute the solution of (6.3) as well as robustifications against model parameter uncertainty, we can apply a simple modification of Algorithm 3 which takes the maximization and the reversed inequality into account. The convergence of this modified algorithm follows in analogy to Theorem 5.8. Based on this theoretical foundation, we now proceed with the solution of the static sub-replication problem.

First we focus on the Black–Scholes model with the same problem data as in the super-replication case presented in Subsection 5.1.3. Hence we consider the sub-replication of an up-and-out call with maturity $T = 1$ year, strike $K = 2750$ and barrier $D = 3300$ based on the spot $S_0 = 2750$, a constant volatility $\sigma_0 = 20\%$, the risk-free rate $r = 5.5\%$ and a dividend yield $\delta = 2.5\%$. In the constant volatility case the sets Θ_1, Θ_2 reduce to $[0, T] \times \{\sigma_0^2\}$ and $[0, D]$ (see Section 5.1).

From a technical point of view, we could now run the modified version of Algorithm 3 with the calls listed in Table 5.1. However, a closer look at the portfolio instruments reveals that the calls in the portfolio are not yet suited for the computation of a sub-replication strategy. The reason for this is graphically visible in Figure 4.3 which illustrates the value of the optimal super-replication portfolio listed in Table 5.3 along the barrier as well as at terminal time. According to Lemma 4.6, the positive portfolio value in the corner (D, T) has to be connected continuously to the value of zero on the barrier. But this implies that the portfolio must have positive value on the barrier shortly before T. However, a positive portfolio value on the barrier violates the first inequality in problem (6.3) and hence the definition of a sub-replicating strategy. Obviously, the sub-replication conditions can only be fulfilled if the portfolio provides a non-positive value in the corner (D, T). Regarding the calls listed in Table 5.1, this can only be achieved by choosing a non-positive quantity of the call with the same strike and maturity as the barrier option. This of course results in extremely bad lower bounds for the fair value of the barrier option. The solution of this problem is to include a call in the hedge portfolio which allows the portfolio value to drop to zero in the corner (D, T) without choosing impractical quantities of the other calls. In our case, this is realized by replacing the call C^2 with strike $K_2 = 3300$ and maturity $T_2 = 1$ by a call with the same maturity, but a lower strike of 3250.

Solving the corresponding sub-replication problem in the Black–Scholes model leads to the optimal portfolio weights listed in Table 6.1. The cost of the optimal sub-replication portfolio is 1.11%, which is only two basis points cheaper than the fair value $\pi(C_{\text{uo}}) = 1.13\%$ of the up-and-out call. Combined with the super-replication price 1.14% (see Subsection 5.1.3), this leads to a tight price interval $[1.11\%, 1.14\%]$ for the value of the up-and-out call. The remarkably small diameter of this interval of only three basis points underlines the quality of the approximation of the barrier option by the set of standard calls $C^1, \dots, C^{10}$.

	C^1	C^2	C^3	C^4	C^5	C^6	C^7	C^8	C^9	C^{10}
T_i	1.00	1.00	1.00	1.00	1.00	0.75	0.75	0.75	0.50	0.50
K_i	2750	3250	3350	3450	3600	3300	3400	3600	3300	3500
α_i	1.00	−11.00	10.87	0.41	−0.82	−0.17	0.38	−0.52	0.04	−0.03

Table 6.1: Optimal sub-replication portfolio weights α_i

Comparing the portfolio weights of the sub-replication portfolio with those of the super-replication portfolio listed in Table 5.3 shows similarities in the sign of the calendar spread part of the portfolio (calls $C^6, \dots, C^{10}$). However, the strike spread component significantly differs from the super-replication case. This difference is illustrated graphically in Figure 6.1. Obviously, the super-replication portfolio perfectly fits the value of the barrier option at terminal time up to the barrier $D = 3300$. At this point, the intrinsic value of the portfolio is exactly $D - K = 550 > 0$ and remains positive within a small interval above the barrier. Hence, at the time of a barrier hit shortly before T, it is most likely that the stock price at terminal time will fall into an interval where the portfolio has a large positive value. This in turn implies a positive portfolio value on the barrier shortly before T and hence illustrates that the call with strike 3300 and maturity of one year is well-chosen for the super-replication case. In contrast to this, the second plot in Figure 6.1 shows that the sub-replication portfolio perfectly fits the barrier option for stock prices up to 3250, but then the new call C^2 with strike

$K_2 = 3250$ allows the portfolio value to drop to exactly zero at $D = 3300$. This guarantees that the portfolio value continuously connects to zero without violating the sub-replication conditions (6.3). But note that besides this conceptual difference the super- and sub-replication portfolio are similar in that they reflect the barrier option payoff above D.

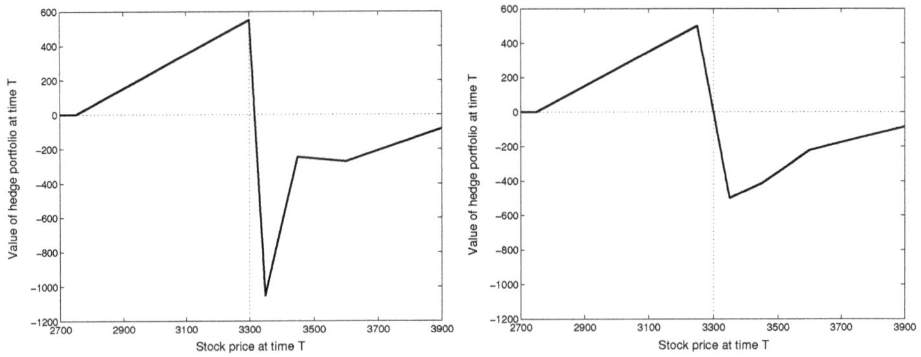

Figure 6.1: Strike spread component of the super-replication portfolio (left) and the sub-replication portfolio (right)

The results presented so far were limited to the case of a constant volatility parameter σ_0. However, as presented in Chapter 5, model parameter uncertainty can lead to extreme hedging losses if it is not taken into account. In the super-replication setting, Subsection 5.1.3 showed that the Black–Scholes hedging strategy can be robustified against changes of the volatility parameter with surprisingly low cost. This lead to robust static upper bounds for the fair value of the barrier option. In analogy it is possible to define robustified versions of problem (6.3) by enforcing the sub-replication property for a given set of volatility parameters $\sigma \in [\sigma_{\min}, \sigma_{\max}]$. As before, existence and convergence results similarly carry over from the robust super- to the robust sub-replication setting.

The question remains how the lower sub-replication bounds change if we gradually increase the degree of robustness. By intuition, the cost of the sub-replication portfolio has to decrease if we add more and more constraints. Running a modified version of Algorithm 3 we obtain the optimal robust sub-replication prices presented in Table 6.2. For completeness we also list the super-replication prices taken from Table 5.7. Starting with the non-robust sub-replication cost of 1.11%, the cost decreases to 1.08% as we increase the size of the set $[\sigma_{\min}, \sigma_{\max}]$. Note that the sub-replication cost does not decrease as quickly as the cost of the super-replication portfolio increases. In combination, the super- and sub-replication prices form robust static bounds for the fair value of the barrier option and hence quantify the volatility risk for this particular option. Surprisingly, the bounds we obtain are very tight.

In analogy to the Black–Scholes case it is possible to compute robust static sub-

Lower bound $\sigma_{\min}$	20%	15%	10%	5%	0%
Upper bound $\sigma_{\max}$	20%	25%	30%	50%	100%
Sub-Replication Price	1.11%	1.10%	1.09%	1.08%	1.08%
Super-Replication Price	1.14%	1.19%	1.27%	1.34%	1.48%

Table 6.2: Robust portfolio bounds for the Black–Scholes model

replication strategies in Heston's stochastic volatility model. To make the findings comparable to the super-replication results (see Table 5.15), we choose the same problem data as in Subsection 5.2.3. Applying a modification of Algorithm 3 to the robust static sub-replication problem and the set of calls listed in Table 6.1 leads to the results presented in Table 6.3.

Size Δ	5%	10%	15%	20%	30%	Non-unif.
$\sqrt{Y_{\min}} = \sqrt{\theta_{\min}}$	17.5%	15.0%	12.5%	10.0%	10.0%	17.5%
$\sqrt{Y_{\max}} = \sqrt{\theta_{\max}}$	22.5%	25.0%	27.5%	30.0%	40.0%	42.5%
$\xi_{\min}$	17.5%	15.0%	12.5%	10.0%	10.0%	45.0%
$\xi_{\max}$	22.5%	25.0%	27.5%	30.0%	40.0%	105.0%
$\kappa_{\min}$	147.5%	145.0%	142.5%	140.0%	130.0%	130.0%
$\kappa_{\max}$	152.5%	155.0%	157.5%	160.0%	160.0%	150.0%
$\rho_{\min}$	−52.5%	−55.0%	−57.5%	−60.0%	−70.0%	−95.0%
$\rho_{\max}$	−47.5%	−45.0%	−42.5%	−40.0%	−40.0%	−55.0%
Sub-Rep. Price	1.45%	1.41%	1.38%	1.35%	1.34%	1.59%
Super-Rep. Price	1.62%	1.73%	1.84%	1.95%	2.07%	2.08%

Table 6.3: Robust portfolio bounds for Heston's stochastic volatility model

Regarding the fair value of the up-and-out call (1.60%, see Subsection 5.2.3), the bounds are not as tight as in the Black–Scholes case. However, compared to the achieved degree of robustness, the sub- and super-replication prices are still very close together. In particular for the non-uniform parameter intervals capturing the parameter tracks in Figure 5.7 the spread of 49 basis points is surprisingly small.

6.2 Robustification against Jumps

So far we showed that the static super-replication approach can be robustified against various sorts of uncertainties. Volatility shocks and skew risk have already been taken into account in Chapter 5 by appropriate modifications of the semi-infinite optimization problem. The numerical results underlined that the robustification against changes of the volatility surface can be gained by surprisingly low cost. Furthermore, the obtained robust super-replication price for the barrier option is of practical relevance, because it is also an upper bound for the price of dynamic hedging strategies which are faced with skew risk.

However, even though various risk factors have been taken into account, the static portfolio derived in Chapter 5 might still suffer from losses if the hedging strategy is applied in practice. The main reason for this is that all the previous results assumed the possibility to liquidate the calls in the portfolio exactly at the time of the barrier hit, that is if $S_\tau = D$ for $\tau \leq T$. But while this is the theoretically correct formulation of the static hedging problem in a financial market model with continuous sample paths, the assumption will be violated almost surely in practice. In the real world, there might not be enough liquidity in the market to immediately sell the portfolio on the barrier or the stock price may simply jump over D without touching it. Correspondingly, we need to analyze the value of our hedge portfolio if the stock price at the liquidation time equals $s > D$.

Without loss of generality we focus in the following on the special case (5.1) of super-replicating an up-and-out call in the Black–Scholes model which is described by the semi-infinite optimization problem

$$\begin{aligned}
& \min_{\alpha \in \mathbb{R}^{n+1}} \alpha_0 B_0 + \sum_{i=1}^{n} \alpha_i C^i(0, S_0, \sigma_0^2) \\
\text{s.t.} \quad & \alpha_0 B_t + \sum_{C^i \in \mathcal{C}, T_i \geq t} \alpha_i C^i(t, D, \sigma_0^2) \geq 0 \quad \forall\, t \in [0, T] \qquad (6.4) \\
& \alpha_0 B_T + \sum_{C^i \in \mathcal{C}, T_i = T} \alpha_i (s - K_i)^+ \geq (s - K)^+ \quad \forall\, s \in [0, D].
\end{aligned}$$

We pick up the problem data which was also the basis for the computations in Subsection 5.1.3, i.e. a risk-free rate $r = 5.5\%$, a dividend yield $\delta = 2.5\%$, the volatility $\sigma_0 = 20\%$, an initial stock price $S_0 = 2750$, and an up-and-out call C_{uo} with strike $K = 2750$, barrier $D = 3300$ and maturity $T = 1$. For this data we computed the optimal super-replication portfolio listed in Table 5.3.

By definition this solution of problem (6.4) guarantees the super-replication property on the barrier D for any time $t \in [0, T]$. This is also graphically visible in the left plot of Figure 6.2 which shows a zoom-in of the portfolio value on the barrier. But what happens if the stock price is above the barrier at the time we unwind our hedge portfolio? The answer is given in the right plot of Figure 6.2 which graphs the portfolio value for a variety of time- and stock price combinations. Obviously, the super-replication property is limited to an exact liquidation on the barrier. In particular the plot reveals a potential loss of more than 1.5% which is unacceptable compared to the fair value $\pi(C_{\text{uo}}) = 1.13\%$ of the barrier option. By inspection, among the presented cases the maximum loss occurs for a stock price of $s = 3310$ close to maturity of the barrier option.

On first sight these large losses seem surprising, but a brief look at the strike spread component of the static super-replication portfolio (see Figure 6.1) delivers an explanation. The reflection of the call with the same strike and maturity as the barrier option

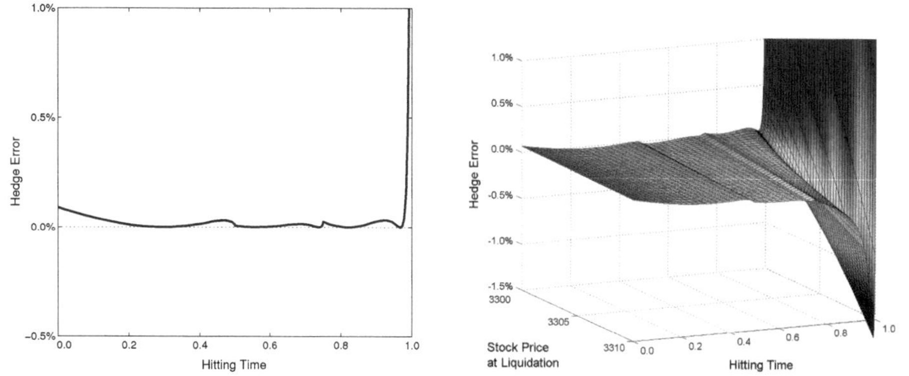

Figure 6.2: Hedge error on the barrier $D = 3300$ (left) and for a variety of stock prices above the barrier (right)

leads to extremely negative portfolio values at maturity T if the stock price is above the barrier. Consequently, if we unwind the hedge portfolio above the barrier and close to T, the stock price will most likely end up in an area where the portfolio value is extremely negative. This leads to the sharp negative spike observed in Figure 6.2.

The question is how we can avoid this problem. At first, it might be tempting to consider a jump diffusion model instead of a model with continuous sample paths. If the jump size is described by an unbounded function, the super-replication property then simply means that the portfolio must be worth at least as much as the up-and-out call for all stock prices $s \in [D, \infty)$ above the barrier. But this is a far too conservative requirement and would in fact lead to a hedge portfolio which is similar to the model-independent Brown–Hobson–Rogers superhedge described in Subsection 2.3.4. The condition of no loss for any stock price implies that the payoff at maturity cannot attain a negative value. This is precisely the feature of the Brown–Hobson–Rogers hedge (see Figure 2.5), and also the reason why it is much more expensive than a model-dependent static superhedge (see Figure 6.1).

From a practical point of view it is sufficient to super-replicate the barrier option in a prespecified stock price interval $[D, S_{\max}]$ above the barrier. However, to realize this idea we do not need a jump diffusion model, but instead can incorporate the additional requirement into the description of the semi-infinite optimization problem (6.4). This is precisely the basis of the following definition.

Definition 6.1. Consider the problem of super-replicating an up-and-out call with maturity T, strike K and barrier D in the Black–Scholes model. Further let $S_{\max} : [0, T] \to [D, \infty)$ be a given continuous function and assume that the standard calls $C^1, \dots, C^n$ satisfy $K_i \geq S_{\max}(T_i)$ for $T_i < T$. Then a static super-replication strategy which is robust against jumps up to size $S_{\max}$ is defined as a solution of the optimization problem

$$\min_{\alpha\in\mathbb{R}^{n+1}} \alpha_0 B_0 + \sum_{i=1}^{n} \alpha_i C^i(0, S_0, \sigma_0^2)$$

$$\text{s.t.} \quad \alpha_0 B_t + \sum_{C^i\in\mathcal{C}, T_i\geq t} \alpha_i C^i(t, s, \sigma_0^2) \geq 0 \;\forall\, t \in [0, T],\; s \in [D, S_{\max}(t)] \tag{6.5}$$

$$\alpha_0 B_T + \sum_{C^i\in\mathcal{C}, T_i=T} \alpha_i (s - K_i)^+ \geq (s - K)^+ \quad \forall\, s \in [0, D].$$

In the simplest case $S_{\max} \geq D$ is a constant which results in a uniform robustness against jumps over the whole time interval $[0, T]$. To account for small jumps and liquidation delays it is usually sufficient to consider $S_{\max} = D + \Delta_s$, where Δ_s is in the magnitude of up to one percent of D. This situation is illustrated in Figure 6.3 which compares the super-replication constraints with and without robustness against jumps.

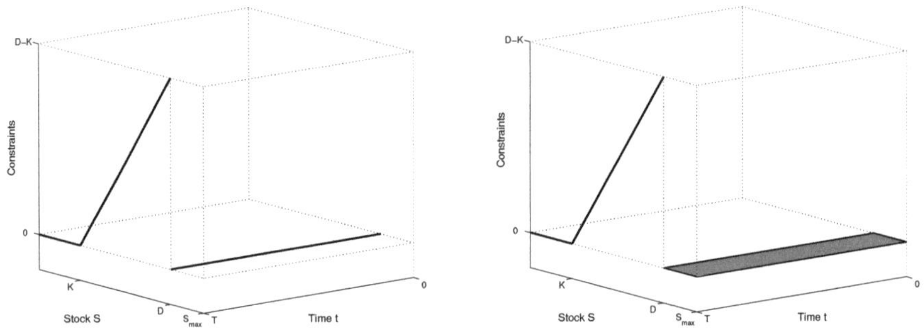

Figure 6.3: Super-replication constraints with (right) and without (left) robustness against jumps

Obviously, the new problem formulation increases the dimension of the semi-infinite parameter space by one. Nevertheless it is still possible to solve problem (6.5) with the algorithms presented in Chapter 5. However, the following theorem shows that it is possible to reduce the dimension of the constraint system to one again by applying the maximum principle for parabolic partial differential equations.

Theorem 6.2. *Under the assumptions of Definition 6.1 the super-replication problem (6.5) with robustness against jumps up to size $S_{\max}$ is equivalent to*

$$\min_{\alpha\in\mathbb{R}^{n+1}} \alpha_0 B_0 + \sum_{i=1}^{n} \alpha_i C^i(0, S_0, \sigma_0^2)$$

$$\text{s.t.} \qquad \alpha_0 B_t + \sum_{C^i \in \mathfrak{C}, T_i \geq t} \alpha_i C^i(t, S_{\max}(t), \sigma_0^2) \geq 0 \quad \forall\, t \in [0, T] \tag{6.6}$$

$$\alpha_0 B_T + \sum_{C^i \in \mathfrak{C}, T_i = T} \alpha_i (s - K_i)^+ \geq (s - K)^+ 1_{[0,D]}(s) \quad \forall\, s \in [0, S_{\max}(T)].$$

Proof. It is sufficient to prove that the feasible sets of problem (6.5) and (6.6) are equal. Clearly, a feasible point of (6.5) is also feasible for (6.6). Therefore, we focus on the question if a feasible point of (6.6) is also feasible for (6.5).

We start by grouping the standard calls $C^1, \ldots, C^n$ into sets $I_1, \ldots, I_r$ with equal maturities $\bar{T}_1 < \ldots < \bar{T}_r = T$. By the superposition principle, the function $c(t,s) := \alpha_0 B_t + \sum_{i \in I_r} \alpha_i C^i(t, s, \sigma_0^2)$ satisfies the Black–Scholes partial differential equation (5.2). But then the maximum principle (see Friedman [42]) implies that the maximum and minimum of c on $\{(t,s) : t \in [\bar{T}_{r-1}, \bar{T}_r],\ s \in [0, S_{\max}(t)]\}$ is attained on the boundaries $[\bar{T}_{r-1}, \bar{T}_r] \times \{0\}$ and $\{(t, S_{\max}(t)) : t \in [\bar{T}_{r-1}, \bar{T}_r]\}$ or on $\{\bar{T}_r\} \times [0, S_{\max}(\bar{T}_r)]$. Hence, if we knew that $c(t,s)$ would be greater or equal to zero on $[\bar{T}_{r-1}, \bar{T}_r] \times \{0\}$, then the constraints

$$c(t, S_{\max}(t)) \geq 0 \ \forall\, t \in [\bar{T}_{r-1}, \bar{T}_r], \ \ c(\bar{T}_r, s) \geq (s - K)^+ 1_{[0,D]}(s) \ \forall\, s \in [0, S_{\max}(\bar{T}_r)]$$

would imply that

$$c(t,s) \geq 0 \ \forall\, t \in [\bar{T}_{r-1}, \bar{T}_r],\ s \in [0, S_{\max}(t)], \ \ c(\bar{T}_r, s) \geq (s - K)^+ \quad \forall\, s \in [0, D],$$

which means that a point satisfying the inequalities of (6.6) for $t \in [\bar{T}_{r-1}, \bar{T}_r]$ also satisfies those of (6.5). The missing link $c(t,0) \geq 0$ for $t \in [\bar{T}_{r-1}, \bar{T}_r]$ is easy to see, because due to (5.2) all calls C^i, $i \in I_r$, satisfy $C^i(t, 0, \sigma_0^2) = 0$. Furthermore, the second constraint of (6.6) implies for $s = 0$ that $\alpha_0 B_T \geq 0$ and hence $\alpha_0 B_t \geq 0$ for all $t \in [0,T]$. This shows that $c(t,0) \geq 0$ and thus implies the equivalence of the constraint systems of (6.6) and (6.5) for $t \in [\bar{T}_{r-1}, \bar{T}_r]$. Proceeding recursively on the time-strips $[\bar{T}_j, \bar{T}_{j+1}]$ proves the equivalence for the whole time-set $[0,T]$. □

The statement of the theorem has a very simple interpretation. The largest hedge errors have to occur on the outer boundary of the constraints of problem (6.5). Hence it is not necessary to consider the whole shaded area in Figure 6.3, but instead one can restrict the super-replication constraints represented by the rectangle to the one-dimensional outer boundaries $\{T\} \times [D, S_{\max}(T)]$ and $\{(t, S_{\max}(t)) : t \in [0,T]\}$. The resulting semi-infinite optimization problem can then be solved with the same algorithms we presented in Chapter 5 – we only have to slightly adapt the constraint coefficients a_1, a_2 and b_2 to the constraints of problem (6.6). Furthermore, the existence and convergence results derived in Chapter 5 carry over in analogy to this new setting.

Based on these findings we proceed with the numerical solution of problem (6.6). For simplicity, we choose $S_{\max} = 3310$ to be a constant function and solve the super-replication problem for the Black–Scholes parameters introduced at the beginning of this section. The calls under consideration and the resulting optimal portfolio weights are presented in Table 6.4. Note that the strikes of the calls C^i are chosen such that $K_i \geq S_{\max}$ for $T_i < T$ and hence satisfy the assumptions of Definition 6.1.

	C^1	C^2	C^3	C^4	C^5	C^6	C^7	C^8	C^9	C^{10}
T_i	1.00	1.00	1.00	1.00	1.00	0.75	0.75	0.75	0.50	0.50
K_i	2750	3300	3350	3450	3600	3350	3400	3600	3350	3500
α_i	1.00	−28.69	32.57	−4.40	0.13	−0.11	0.14	−0.26	0.03	−0.10

Table 6.4: Optimal Portfolio Weights α_i for the case $S_{\max} = 3310$

The cost of the optimal super-replication portfolio is 1.20%, which is only 7 basis points more expensive than the fair value $\pi(C_{\text{uo}}) = 1.13\%$ of the barrier option. Compared to the potentially extreme losses without a robustification against jumps (see Figure 6.2), this additional risk premium seems to be quite low. It should be emphasized, that the computed portfolio guarantees the super-replication property for stock prices up to $S_{\max} = 3310$ at any possible liquidation time $t \in [0, T]$. However, in practice one might be interested in more conservative choices of $S_{\max}$ to protect the portfolio against losses caused by even bigger jumps. Therefore, Table 6.5 lists the cost of optimal portfolios consisting of the calls listed in Table 6.4 for varying $S_{\max}$.

$S_{\max}$	3300	3310	3320	3330	3340	3350
Cost	1.14%	1.20%	1.25%	1.31%	1.37%	1.44%

Table 6.5: Cost of hedge portfolios with varying robustness against jumps

Obviously, the robustification against big jumps does not come for free. But in order to cover liquidity risk or liquidation delays it might be sufficient to buy protection against stock prices up to $S_{\max} = 3320$ or 3330 which requires a surcharge of 10 to 15 basis points. Also note that the cost grows linearly in the robustness parameter $S_{\max}$, an effect which has already been observed for the robustification against model parameter uncertainty. For completeness we list the optimal portfolio weights for the case $S_{\max} = 3350$ in Table 6.6.

	C^1	C^2	C^3	C^4	C^5	C^6	C^7	C^8	C^9	C^{10}
T_i	1.00	1.00	1.00	1.00	1.00	0.75	0.75	0.75	0.50	0.50
K_i	2750	3300	3350	3450	3600	3350	3400	3600	3350	3500
α_i	1.00	−12.00	4.14	9.57	−2.26	−0.01	0.02	−0.11	0.02	−0.08

Table 6.6: Optimal portfolio weights α_i for the case $S_{\max} = 3350$

As a rule of thumb we can conclude that the robustification against small jumps is possible for an acceptable premium. If we combine this result with the robustification against model parameter uncertainty developed in Chapter 5, we obtain very robust static hedge portfolios which offer enough protection to be implemented in practice. In particular the semi-infinite programming approach allows to compute hedging strategies which are robust against volatility shocks, changes of the skew and jumps. All these robustifications can be viewed as components which can be combined arbitrarily. As we also consider calls which are actually traded on the EUREX, the only real assumption that is left is that option markets are liquid enough to set up and unwind the hedge and that traders do not mind to face the huge hedge positions with opposite signs.

If we assume the availability of calls with arbitrary strikes, we can improve the results presented above even further. For example, we can add the calls with strikes 3301 and 3349 and maturity $T = 1$ to the set of hedge instruments listed in Table 6.6. In this case a solution of problem (6.6) with constant $S_{\max} = 3350$ leads to an optimal portfolio with a cost of 1.41%. This only slightly improves the cost 1.44% of the hedge portfolio presented in Table 6.6 and hence demonstrates that an optimal choice of available calls leads to very good results.

However, if we plot the strike spread component of the two separate portfolios we obtain very interesting insights into the structure of the hedge positions. Figure 6.4 reveals that the optimal portfolio consisting of the additional two calls nearly perfectly matches the super-replication constraint of optimization problem (6.6). Above the barrier $D = 3300$, the portfolio value immediately drops to zero and remains there until a stock price of $s = 3349$ is reached. At this point the portfolio makes use of the available call with strike 3349 to slightly shift the portfolio value upwards before reflecting the payoff of the call without knock-out feature. The short spike at $s = 3349$ is needed to compensate the negative portfolio value above $S_{\max} = 3350$ and hence guarantees the super-replication property if we move away from $t = T$. Compared to this, the strike spread component of the portfolio listed in Table 6.6 looks much more regular and achieves the super-replication property by a less rapid drop of the portfolio value above D. Note that the optimization algorithm combined the portfolio positions in such a way that its value in the sum crosses zero at $S_{\max}$. This is also intuitively clear: The payoff of the call without knock-out feature must be reflected above the barrier as soon as possible. If we robustify against jumps, the earliest possibility for this reflection is at $S_{\max}$.

Based on Figure 6.4 the idea arises to guarantee the robustness against jumps in another way. Instead of requesting that the portfolio value remains non-negative in the interval $(D, S_{\max}]$ we might simply ask the portfolio to be greater or equal to $(s-K)^+$ in this interval, too. This leads to the following simple modification of optimization problem (6.6) with constant $S_{\max}$.

$$\min_{\alpha\in\mathbb{R}^{n+1}} \alpha_0 B_0 + \sum_{i=1}^{n} \alpha_i C^i(0, S_0, \sigma_0^2)$$

$$\text{s.t.} \qquad \alpha_0 B_t + \sum_{C^i\in\mathcal{C}, T_i\geq t} \alpha_i C^i(t, S_{\max}, \sigma_0^2) \geq 0 \;\forall\, t \in [0, T] \tag{6.7}$$

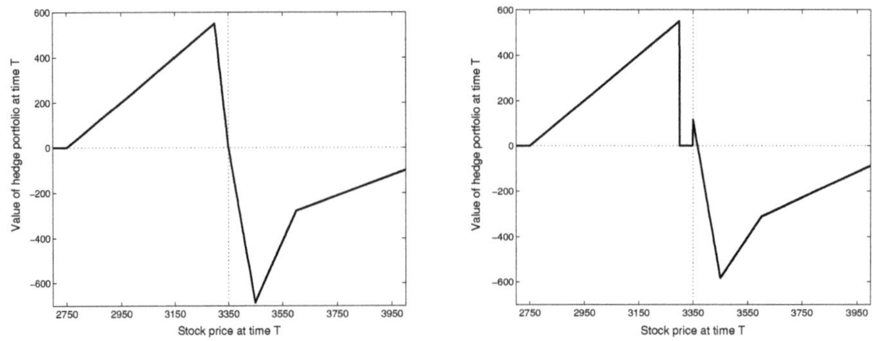

Figure 6.4: Strike spread part of the portfolio listed in Table 6.6 (left) and the optimal portfolio we obtain by adding calls with maturity of one year and strikes 3301, 3349 (right)

$$\alpha_0 B_T + \sum_{C^i \in \mathcal{C}, T_i = T} \alpha_i (s - K_i)^+ \geq (s - K)^+ \; \forall \, s \in [0, S_{\max}].$$

The only difference between problems (6.6) and (6.7) is the missing indicator function which was part of the second constraint. This difference is illustrated graphically in Figure 6.5. Obviously, the small change of the portfolio constraints at terminal time for stock prices in the interval $(D, S_{\max}]$ implies that problem (6.7) actually describes the super-replication of an up-and-out call with shifted barrier $S_{\max} \geq D$. Based on the assumption of Definition 6.1, that is $K_i \geq S_{\max}$ for calls with maturity $T_i < T$, problem (6.7) is also guaranteed to be theoretically equivalent to the stochastic super-replication problem (4.3) with shifted barrier (see Theorem 4.4).

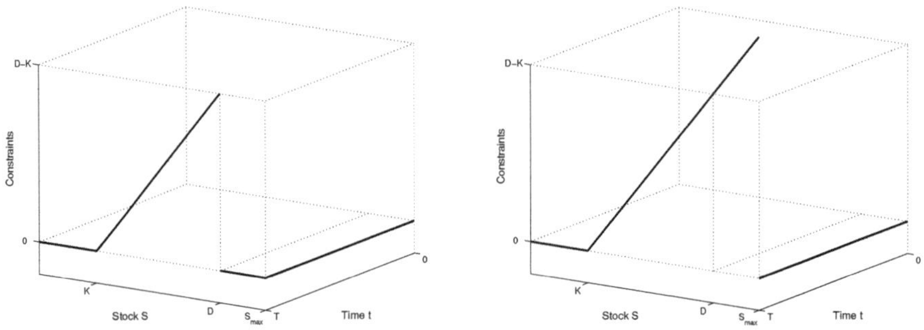

Figure 6.5: Comparison of the super-replication constraints of problem (6.6) with those of problem (6.7)

Note that we exaggerated the length of the interval $[D, S_{\max}]$ in Figure 6.5 to make the differences of the optimization problems visible. Usually, this interval will be tiny compared to the absolute size of D such that the super-replication problem with shifted barrier is actually a very good approximation of problem (6.6). The quality of the approximation is also obvious in Table 6.7 which compares the cost of optimal portfolios solving problems (6.6) and (6.7) for varying $S_{\max}$.

Parameter $S_{\max}$	3300	3310	3320	3330	3340	3350
Cost of α^* solving (6.6)	1.14%	1.20%	1.25%	1.31%	1.37%	1.44%
Cost of α^* solving (6.7)	1.14%	1.27%	1.32%	1.37%	1.42%	1.47%

Table 6.7: Cost of optimal solutions of problems (6.6) and (6.7)

Obviously, the costs of the optimal portfolios are very similar which justifies to interprete problem (6.6) as *moving the barrier* from D to $S_{\max}$. This is a quite common market practice to dynamically hedge barrier options (see Schmock, Shreve and Wystup [79]) and hence underlines the practical relevance of the framework proposed in this section. Finally we mention that problem (6.7) can be solved without any modification of the algorithms developed in Chapter 5.

6.3 Extension to other Barrier Options

The main results of the previous chapters and sections were focused on the special case of hedging an up-and-out call C_{uo} with maturity T, strike K and barrier $D > K$. We chose this particular barrier option, because it has a very interesting discontinuous payoff profile which complicates the dynamic hedging of this option. However, the results derived so far are by no means limited to up-and-out calls. Instead, the theory and algorithms can be generalized to various other sorts of barrier options or even to completely different option classes. In this section we present some examples which show how the semi-infinite programming approach can be applied to the static hedging of other options.

Up-and-In Call The first example we consider is the problem of hedging an up-and-in call. This barrier option is defined by the payoff

$$C_{ui} = (S_T - K)^+ 1_{\{\tau \leq T\}}, \quad \tau := \inf\{t \in [0, T] : S_t \geq D\},$$

and is the simplest modification of the up-and-out call C_{uo}, because the two options are connected via the relation $C_{ui} + C_{uo} = (S_T - K)^+$. If we reformulate this as $C_{ui} = (S_T - K)^+ - C_{uo}$ it is apparent that a hedge portfolio for an up-and-in call can be obtained by going long one call with strike K and maturity T and shorting the hedge portfolio for the up-and-out call. Furthermore, if α^{sup} is the static super-replication strategy for C_{uo} computed in Chapter 5 and α^{sub} denotes the corresponding sub-replication portfolio (see Section 6.1), i.e. $\Pi_T(\alpha^{\text{sup}}) \geq C_{\text{uo}}$ (a.s.) and $\Pi_T(\alpha^{\text{sub}}) \leq C_{\text{uo}}$ (a.s.), then

$$(S_T - K)^+ - \Pi_T(\alpha^{\text{sup}}) \leq C_{\text{ui}} \leq (S_T - K)^+ - \Pi_T(\alpha^{\text{sub}}) \text{ (a.s.)}. \tag{6.8}$$

Thus the super/sub-replication portfolio for C_{uo} leads to a sub/superhedge for C_{ui} which implies that the (robust) static strategies and bounds we computed in Chapter 5 for C_{uo} immediately transfer to the up-and-in call.

For example, in the Black–Scholes model with risk-free rate $r = 5.5\%$, dividend-yield $\delta = 2.5\%$, volatility $\sigma_0 = 20\%$ and spot $S_0 = 2750$, the hedging strategies $\alpha^{\text{sup}}, \alpha^{\text{sub}}$ listed in Tables 5.3 and 6.1 lead to the static sub- and super-replication portfolios presented in Table 6.8 for an up-and-in call with strike $K = 2750$, barrier $D = 3300$ and maturity $T = 1$. Note that $\Pi_0(\alpha^{\text{sup}}) = 1.14\%$, $\Pi_0(\alpha^{\text{sub}}) = 1.11\%$ and that the price of a plain vanilla call with strike 2750 and maturity T is 9.18%. Hence inequalities (6.8) imply the price interval $[9.18\% - 1.14\%, 9.18\% - 1.11\%] = [8.04\%, 8.07\%]$ for the up-and-in call which of course contains the fair value given by $\pi(C_{\text{ui}}) = 9.18\% - \pi(C_{\text{uo}}) = 9.18\% - 1.13\% = 8.05\%$. As in the up-and-out case this interval is very tight.

	C^1	C^2	C^3	C^4	C^5	C^6	C^7	C^8	C^9	C^{10}
T_i	1.00	1.00	1.00	1.00	1.00	0.75	0.75	0.75	0.50	0.50
K_i	3250	3300	3350	3450	3600	3300	3400	3600	3300	3500
α_i	0.00	33.02	−40.06	8.20	−0.79	0.04	−0.04	0.25	−0.02	0.09
β_i	11.00	0.00	−10.87	−0.41	0.82	0.17	−0.38	0.52	−0.04	0.03

Table 6.8: Optimal subhedge α and superhedge β for an up-and-in call

Regarding the practical implementation of α, β we mention that due to the relation $C_{\text{ui}} = (S_T - K)^+ - C_{\text{uo}}$ the hedging strategies are set up by buying the portfolio α, β at time $t = 0$. At the time $\tau \leq T$ of a barrier hit, we unwind the hedge portfolio and buy one standard call with the same strike and maturity as the barrier option.

Down-and-Out/In Calls Now we consider the case of statically hedging a down-and-out call with barrier $D < S_0$, strike $K > D$ and maturity T, whose payoff is defined by

$$C_{\text{do}} = (S_T - K)^+ 1_{\{\tau > T\}}, \quad \tau := \inf\{t \in [0, T] : S_t \leq D\}.$$

In contrast to the up-and-out call, this barrier option has a continuous payoff in the sense of no jump at the point (T, D). However, the static super-replication of C_{do} is still a special case of the general optimization problem (3.1). Hence Theorem 3.3 assures the existence of a solution of the hedging problem if the set of static super-replicating strategies is nonempty and the model is arbitrage-free in the set of knock-out trading strategies introduced in Definition 2.2.

Thus we could proceed immediately with the numerical solution of the stochastic optimization problem (3.1) by applying the Monte Carlo-based Algorithm 1 to this particular problem. However, as in the case of the up-and-out call this method has the drawback of requiring a lot of computation time. In Section 4.1 we were able to reduce the computational complexity for C_{uo} by eliminating the stochastic nature from the static hedging problem. This was achieved by reducing the number of hedge instruments expiring before T to calls with strikes at or above the barrier. These calls have the property to expire worthless as long as the barrier has not been hit. In analogy, for the down-and-out case puts with strikes at or below the barrier have the property to expire worthless as long as the stock price does not touch or cross the barrier. This motivates the following assumption.

Assumption 6.3. Let the set of standard options $\mathcal{O} = \{C^1, \dots, C^n\}$ which are part of the financial market model defined in Assumption 2.1 consist of calls and puts with maturities $T_i \le T$ and strikes K_i satisfying $K_i \ne K_j$ for $T_i = T_j, i \ne j$. Assume for $T_i < T$ that option C^i is a put with strike $K_i \le D$.

If we further assume that the financial market model is described by the general stochastic volatility model defined in Assumption 4.1, then based on Assumption 6.3 one can prove in analogy to Theorem 4.4 that the stochastic super-replication problem (3.1) is equivalent to the semi-infinite optimization problem

$$\begin{aligned} &\min_{\alpha \in \mathbb{R}^{n+1}} \ \alpha_0 B_0 + \textstyle\sum_{i=1}^{n} \alpha_i C^i(0, S_0, Y_0) \\ \text{s.t.} \quad &\alpha_0 B_t + \textstyle\sum_{C^i \in \mathcal{O}, T_i \ge t} \alpha_i C^i(t, D, y) \ge 0 \quad \forall\, (t, y) \in \Theta_1 \qquad (6.9) \\ &\alpha_0 B_T + \textstyle\sum_{C^i \in \mathcal{O}, T_i = T} \alpha_i C^i(T, s, y) \ge (s - K)^+ \quad \forall\, s \in \Theta_2, \end{aligned}$$

where Θ_1 describes the potential time-variance combinations for which the barrier may be hit and Θ_2 consists of the stock prices that can be attained without a barrier hit. For example, in the Black–Scholes model Θ_1 equals $[0, T] \times \{\sigma_0^2\}$ and $\Theta_2 = [D, \infty)$. Note that Θ_2 differs from the up-and-out call case and that the sum over all $C^i \in \mathcal{O}$ with $T_i = T$ in general consists of calls and puts with payoff $C^i(T, s, y) = (s - K_i)^+$ or $C^i(T, s, y) = (K_i - s)^+$, respectively.

In analogy to the derivations in Chapter 5 we can further robustify problem (6.9) against model parameter uncertainty, prove the existence of solutions and convergence of appropriately modified algorithms which are much faster than the Monte Carlo-based method. By numerically solving the robust static super- or sub-replication problem we can compute static bounds which quantify the risk of a changing volatility surface for the down-and-out call. Of course it is also possible to incorporate an additional robustness against jumps as presented in Section 6.2. Further note that the case of a down-and-in call $C_{\mathrm{di}} = (S_T - K)^+ 1_{\{\tau \le T\}}$ can be treated in analogy to the up-and-in call by exploiting the relation $C_{\mathrm{do}} + C_{\mathrm{di}} = (S_T - K)^+$.

Down/Up-and-Out/In Puts In the derivations above there was nothing special about barrier options with a call-based payoff at maturity. We can equally well repeat the arguments for down/up-and-out/in puts. For example, the analog to the up-and-out call is given by the down-and-out put with discontinuous payoff

$$P_{\mathrm{do}} = (K - S_T)^+ 1_{\{\tau > T\}}, \quad \tau := \inf\{t \in [0, T] : S_t \leq D\},$$

where $D < S_0$ denotes the barrier and $K > D$ is the strike. Similar to the down-and-out call, the super-replication problem for P_{do} is a special case of optimization problem (3.1) such that the theory developed in Chapter 3 applies. If we further restrict the set of hedge instruments to those listed in Assumption 6.3 and assume that our market is driven by the stochastic volatility model defined in Assumption 4.1, then the semi-infinite representation of the super-replication problem for P_{do} is given by

$$\begin{aligned} &\min_{\alpha \in \mathbb{R}^{n+1}} \alpha_0 B_0 + \sum_{i=1}^{n} \alpha_i C^i(0, S_0, Y_0) \\ \text{s.t.} \quad &\alpha_0 B_t + \sum_{C^i \in \mathcal{O}, T_i \geq t} \alpha_i C^i(t, D, y) \geq 0 \quad \forall\, (t, y) \in \Theta_1 \qquad (6.10) \\ &\alpha_0 B_T + \sum_{C^i \in \mathcal{O}, T_i = T} \alpha_i C^i(T, s, y) \geq (K - s)^+ \quad \forall\, s \in \Theta_2. \end{aligned}$$

For this problem a set of suitable hedge instruments is given by the standard put with the same strike and maturity as the barrier option, several puts with maturity T and strikes below the barrier and some puts with expiry $T_i < T$ and strikes $K_i \leq D$. Of course we can again robustify problem (6.10) against changes of the volatility surface, stock price jumps or consider the sub-replication of P_{do}. As before, the theory and algorithms developed in previous sections and chapters also apply here.

Note that a down-and-in put $P_{\mathrm{di}} = (K - S_T)^+ 1_{\{\tau \leq T\}}$ can be treated based on the equation $P_{\mathrm{do}} + P_{\mathrm{di}} = (K - S_T)^+$. Furthermore, we can set up semi-infinite programming problems for up-and-out puts and up-and-in puts with barrier $D > S_0$, if the hedge instruments expiring before T are restricted to calls with strikes $K_i \geq D$ that expire worthless as long as the barrier has not been hit.

Double Knock-Out Options It is not surprising that the semi-infinite programming framework is also applicable to the case of double barrier options. Given two barriers $D_1 < S_0 < D_2$ and two stopping times $\tau_1 := \inf\{t \in [0, T] : S_t \leq D_1\}$, $\tau_2 := \inf\{t \in [0, T] : S_t \geq D_2\}$, the payoff of a double knock-out call with strike K and maturity T is for example defined by

$$C_{\mathrm{dk}} = (S_T - K)^+ 1_{\{\tau > T\}}, \quad \tau := \min\{\tau_1, \tau_2\}.$$

Again, the super-replication of C_{dk} is a special case of problem (3.1) for which Theorem 3.3 guarantees the existence of a solution under suitable assumptions. To eliminate the stochastic nature we can restrict those hedge instruments in the set $\mathcal{O} =$

$\{C^1, \ldots, C^n\}$ expiring before T to calls with strikes $K_i \geq D_2$ and puts with $K_i \leq D_1$. This guarantees that both types of contracts expire worthless as long as none of the barriers have been hit. Without going into the details it is intuitively clear that in the Black–Scholes model with constant volatility σ_0, the stochastic super-replication of the double knock-out call is equivalent to the semi-infinite optimization problem

$$\min_{\alpha \in \mathbb{R}^{n+1}} \alpha_0 B_0 + \sum_{i=1}^{n} \alpha_i C^i(0, S_0, \sigma_0^2)$$

$$\text{s.t.} \quad \alpha_0 B_t + \sum_{C^i \in \mathcal{O}, T_i \geq t} \alpha_i C^i(t, D_1, \sigma_0^2) \geq 0 \quad \forall\, t \in [0, T]$$

$$\alpha_0 B_t + \sum_{C^i \in \mathcal{O}, T_i \geq t} \alpha_i C^i(t, D_2, \sigma_0^2) \geq 0 \quad \forall\, t \in [0, T]$$

$$\alpha_0 B_T + \sum_{C^i \in \mathcal{O}, T_i = T} \alpha_i C^i(T, s, y) \geq (s - K)^+ \quad \forall\, s \in [D_1, D_2].$$

Similar variants of this problem can be derived for other models like the general stochastic volatility model described in Assumption 4.1. As calls and puts with strikes above D_2 and below D_1 are part of the hedge portfolio and hence form a big subset of the volatility surface, the robustification against volatility shocks and skew risk is even more important for double knock-out options than in the single barrier case. However, the optimization framework allows to easily incorporate this robustness as well as the derivation of existence and convergence results for algorithms taking the double barrier feature into account. Furthermore we can consider stock price jumps over both barriers or analyze the sub-replication problem.

Non-Zero Rebates The knock-out barrier options we considered up to this point expired worthless in the case of a barrier hit. However, some barrier contracts instead pay a predefined rebate if the option knocks out. This payment may be delivered at the time of the barrier hit or at maturity T. Note that both cases can be dealt with by a time-dependent rebate function $R : [0, T] \to [0, \infty)$, where $R(t)$ represents the payment if the barrier is breached at time t. If a constant rebate $\bar{R}$ is paid at the hitting time, then $R(t) = \bar{R}\ \forall\, t$, whereas a fixed rebate $\bar{R}$ paid at maturity T can be incorporated by choosing $R(t) = \bar{R}e^{-r(T-t)}$ if the risk-free rate r is constant.

Still the super-replication of a knock-out barrier option with rebate feature is a special case of the stochastic optimization problem (3.1) and can be reformulated as a semi-infinite optimization problem as before. The only thing that changes is the right hand side of the inequality enforcing the super-replication at the time of a barrier hit. For example, for a down-and-out put with rebate function R, the problem (6.10) describing the zero rebates case simply changes to

$$\min_{\alpha \in \mathbb{R}^{n+1}} \alpha_0 B_0 + \sum_{i=1}^{n} \alpha_i C^i(0, S_0, Y_0)$$

$$\text{s.t.} \quad \alpha_0 B_t + \sum_{C^i \in \mathcal{O}, T_i \geq t} \alpha_i C^i(t, D, y) \geq R(t) \quad \forall\, (t, y) \in \Theta_1$$

$$\alpha_0 B_T + \sum_{C^i \in \mathcal{O}, T_i = T} \alpha_i C^i(T, s, y) \geq (K - s)^+ \quad \forall\, s \in \Theta_2.$$

Accordingly, the algorithms developed in previous chapters can be applied by simply changing the right hand side of the super-replication inequality. Furthermore, existence and convergence results also transfer to the non-zero rebates case.

Note that in analogy also knock-in options can be considered which pay a rebate R if the barrier has not been hit until maturity T. Finally we mention that so-called American digital calls and puts can be viewed as knock-in barrier options with non-zero rebate but zero payoff otherwise. Hence the proposed framework is also applicable to the robust static hedging of these products.

Non-Constant Barriers The next generalization we aim at affects the shape of the barrier which was assumed to be constant in all previous problems. Without loss of generality we now consider an up-and-out call with zero rebates and continuous non-constant barrier function $D : [0, T] \to (0, \infty)$, $D(0) > S_0$, which is defined by the payoff

$$C_{\text{uo}} = (S_T - K)^+ 1_{\{\tau > T\}}, \quad \tau := \inf\{t \in [0, T] : S_t \geq D(t)\}.$$

Again τ is a stopping time and hence the static super-replication problem for C_{uo} is covered by the general theory presented in Chapter 3. Further we can show for the market model defined in Assumption 4.1 and the hedge instruments satisfying Assumption 4.2 for $D(t)$ instead of D that the stochastic hedging problem is equivalent to

$$\min_{\alpha \in \mathbb{R}^{n+1}} \alpha_0 B_0 + \sum_{i=1}^{n} \alpha_i C^i(0, S_0, Y_0)$$

$$\text{s.t.} \quad \alpha_0 B_t + \sum_{C^i \in \mathcal{C}, T_i \geq t} \alpha_i C^i(t, D(t), y) \geq 0 \quad \forall\, (t, y) \in \Theta_1 \tag{6.11}$$

$$\alpha_0 B_T + \sum_{C^i \in \mathcal{C}, T_i = T} \alpha_i (s - K_i)^+ \geq (s - K)^+ \quad \forall\, s \in \Theta_2,$$

where Θ_1 now describes the time-variance combinations (t, y) for which the barrier $D(t)$ might be hit by the stock price at time t and Θ_2 denotes the stock-prices that are attainable without a barrier hit. For example we obtain for the Black–Scholes model $\Theta_1 = [0, T] \times \{\sigma_0^2\}$ and $\Theta_2 = [0, D(T)]$. The main difference between problem (6.11) and the constant barrier case (4.4) is hence that we replaced D by $D(t)$ and obtain potentially different sets Θ_1, Θ_2. Apart from that the problem can be robustified as before and the usual existence and convergence theory applies.

Discrete Barriers Instead of continuously monitored barrier options we now consider knock-out options with discrete barrier. For example an up-and-out call with discrete barrier $D > 0$ pays off $(S_T - K)^+$ at maturity T, provided the stock price is below D at prespecified times $0 < t_1 < \ldots < t_N \leq T$. If we define $\tau := \min\{t \in \{t_1, \ldots, t_N\} : S_t \geq D\}$, $\min \emptyset := +\infty$, then the option payoff is hence given by $C_{\text{uo}} = (S_T - K)^+ 1_{\{\tau > T\}}$. Once again the super-replication problem is a special case of the general stochastic hedging problem (3.1).

However, this time the equivalent semi-infinite super-replication problem looks very different, because the stock can attain values above D between the time slices t_j without leading to a knock-out of the barrier option. Instead of the knock-out region $[0, T] \times \{D\}$ for the case of a continuous barrier option we now identify the barrier sets $\{t_j\} \times [D, \infty)$, $j = 1, \ldots, N$, to be the crucial time/stock price-combinations for which the option knocks out. Consequently, the super-replication property has to be preserved on these sets. For the Black–Scholes model this leads to the semi-infinite optimization problem

$$\min_{\alpha \in \mathbb{R}^{n+1}} \alpha_0 B_0 + \sum_{i=1}^{n} \alpha_i C^i(0, S_0, \sigma_0^2)$$

$$\text{s.t.} \quad \alpha_0 B_{t_j} + \sum_{C^i \in \mathcal{C}, T_i \geq t_j} \alpha_i C^i(t_j, s, \sigma_0^2) \geq 0 \quad \forall\, s \in [D, \infty),\ j = 1, \ldots, N$$

$$\alpha_0 B_T + \sum_{C^i \in \mathcal{C}, T_i = T} \alpha_i (s - K_i)^+ \geq (s - K)^+ \quad \forall\, s \in [0, \infty),$$

if the set of hedge instruments $C^1, \ldots, C^n$ only consists of calls with maturities $T_i \in \{t_1, \ldots, t_N, T\}$ and strikes $K_i \geq D$ in case $T_i < T$. Obviously, the number of constraints has increased significantly in contrast to problem (4.4) of hedging an up-and-out call with continuous barrier. However, note that each of the sets $\{t_j\} \times [D, \infty)$ is a one-dimensional semi-infinite parameter set such that the slack minimization for each of these time slices is not too hard to carry out. Hence a solution of the super-replication problem can be numerically computed by straight-forward generalizations of the semi-infinite algorithms presented in Chapter 5.

Furthermore, robustifications against model parameter uncertainty can easily be incorporated into the description of the optimization problem. Combined with an analogous sub-replication strategy, we also obtain robust static bounds for knock-out options with discrete barriers. But in contrast to the continuous barrier case an additional protection against jumps is not necessary, because the super-replication property by definition is preserved for all stock prices above the barrier.

Other Option Classes The previous generalizations of the static super-replication problem demonstrated the flexibility of the semi-infinite programming approach for various other sorts of barrier options. But in fact the framework we developed is a very general hedging concept rather than a method which is particularly tailored to barrier options.

In the sense of Joshi [55], barrier options are one example of an exotic option which can be *weakly replicated*, that is by setting up a finite portfolio of calls and puts today which may be sold before their own expiries. But while Joshi and the static hedging approaches in the literature assume that the future smile is a *known* deterministic function, the approach we presented allows for an infinite number of *unknown* future smile scenarios which are parametrized by a given financial market model. The static hedging strategy we compute super-replicates the target option in any of the smile scenarios and hence leads to an infinite number of constraints in the optimization problem. The resulting hedging strategies are robust with respect to changes of the volatility surface and can be computed via appropriate semi-infinite optimization techniques. Even if option markets may not be liquid enough nowadays to implement the static hedging strategies in practice, the bounds we obtain are upper bounds for the risk of changes in the volatility surface and thus are of practical relevance for dynamic trading strategies, too.

This approach can be applied to arbitrary exotic options which can be weakly replicated. Moreover, the idea even seems to extend to claims (for example Asian options) which are hedgeable by a *feeble static* trading strategy, i.e. a portfolio consisting of a finite number of calls and puts which can be traded at a finite number of times. For these options it might be possible to recursively compute robust super-replicating strategies, starting with the payoff at maturity T and then proceeding backwards in time. But further research is needed in this direction to support this conjecture.

7 Avoiding Model Errors

Throughout the previous chapters we illustrated the intuitiveness and flexibility of the static super-replication approach and in particular its semi-infinite representation by several robustifications and extensions. From a financial point of view the robustification against model parameter uncertainty was the most important step, because it quantifies and eliminates the skew risk from static hedge portfolios for barrier options. The idea behind this approach was to guarantee the super-replication property for a given set of model parameters corresponding to an infinite number of potential future scenarios of the volatility surface. This offers perfect protection against changing model parameters within the given financial market model, but as soon as the model prices differ from those observed in reality, the hedge performance might suffer due to these so-called *model errors*.

Advanced financial market models like Heston's stochastic volatility model already calibrate fairly well to given volatility surfaces. This means in terms of the call option prices that the calibrated model prices are already close to the market prices (usually the average calibration error is in the magnitude of a few basis points). As an implication, we can expect that the effect of model errors is of minor importance in comparison to model parameter uncertainty. However, the question still persists if we can additionally robustify the static hedge portfolio to prevent hedging losses.

In this chapter we present a method developed by Leibfritz and Maruhn in [64] to eliminate the model error risk from the hedging strategy by allowing the call option prices to deviate in a small neighborhood of the model option prices. These deviations are mathematically described as ellipsoidal uncertainty sets around the asset prices and result in a linear semi-infinite optimization problem with a typically 15 to 30-dimensional parameter space of the semi-infinite constraints. Clearly, the high dimension of this parameter space makes the numerical solution of the problem very hard. However, by using equivalent transformations, we can reduce the dimension of the parameter space significantly. The price for this reduction is a nonlinearity in the form of second order cone constraints. By employing suitable semi-infinite optimization results, we can prove convergence of an iterative method successively solving second order cone programs and nonlinear programming problems to compute the worst case constraint violation for each iterate. In addition we are able to reformulate this iterative procedure as a successive solution of semidefinite programming problems and nonlinear semidefinite programs (NSDPs). In particular our derivations show that NSDPs naturally arise as subproblems of robustified linear semi-infinite programming problems to compute the constraint violation of the iterates.

Nonlinear semidefinite programs also often appear in the design of static output feedback control laws for linear time-invariant control systems (see e.g. [62], [66]). Finding a solution of these non-convex NSDPs is a difficult task, particularly if the dimension of the problem is large. This may be one reason why general "off the shelf solvers" for NSDPs are unavailable. To the best of the knowledge of the author, there only exist some specialized solvers for particular NSDPs (see for example SLMSDP [56], PENBMI [61], IPCTR [65], SSDP [36]). On the other hand, for the solution of linear SDPs and SOCPs a lot of solvers are freely available over the internet (e.g. DSDP [10], SeDuMi [82], SDPT3 [84] and many more). Due to this gap, it is necessary to develop, test and analyze solvers for more general NSDPs. The financial market application presented in this chapter provides an additional example why the development of general NSDP solvers is needed.

The chapter is organized as follows. In Section 7.1 we pick up the semi-infinite programming formulation of the static hedging approach and add robustness to the problem by taking possible model as well as implementation errors into account. Furthermore, we briefly sketch the numerical method to solve the problem and derive the convergence of the iterative procedure. In Section 7.2 we show how the robust optimization problem can be solved by a sequence of SDPs and NSDPs. In particular we prove that a nonlinear second order cone problem is equivalent to the minimization of the minimal eigenvalue of a matrix function, where this matrix depends in a very nonlinear fashion on the parameters. Then, by using duality arguments, the problem of minimizing the minimal eigenvalue of this matrix function will be reformulated as a nonlinear minimization problem with SDP constraints. Finally, Section 7.3 presents numerical results which prove that the proposed algorithm can successfully be applied to improve the robustness of portfolio strategies.

7.1 Robustification and Numerical Solution

The starting point of our analysis is the semi-infinite formulation of the static super-replication problem for stochastic volatility models, extended by the robustification against model parameter uncertainty presented in Section 5.2. This leads to the deterministic optimization problem (5.9), where the model parameters p are allowed to vary within the compact uncertainty set U while still preserving the super-replication property. Without loss of generality we restrict ourselves in the following to the case of Heston's stochastic volatility model. In this model the super-replication of an up-and-out call is described by

$$\min_{\alpha \in \mathbb{R}^{n+1}} \ \alpha_0 B_0 + \sum_{i=1}^{n} \alpha_i C^i(0, S_0, Y_0, p_0)$$

$$\text{s.t.} \quad \alpha_0 B_t + \sum_{C^i \in \mathfrak{C}, T_i \geq t} \alpha_i C^i(t, D, y, p) \geq 0 \ \forall \ (t, y, p) \in [0, T] \times [Y_{\min}, Y_{\max}] \times U$$

$$\alpha_0 B_T + \sum_{C^i \in \mathfrak{C}, T_i = T} \alpha_i (s - K_i)^+ \geq (s - K)^+ \quad \forall \ s \in [0, D]$$

$$\alpha_i^{\mathrm{lb}} \le \alpha_i \le \alpha_i^{\mathrm{ub}},\ i = 0, \dots, n,$$

where we added box constraints $\alpha_i^{\mathrm{lb}} \le 0$, $\alpha_i^{\mathrm{ub}} \ge 1$, and replaced the set $\Theta_1 = [0, T] \times [0, \infty)$ by a robustness interval $\Gamma_1 = [0, T] \times [Y_{\min}, Y_{\max}]$ as suggested by Figure 5.7. The Heston model parameters $p \in \mathbb{R}^4$ are given by the mean reversion speed κ, the long term mean of the variance θ, the volatility of volatility ξ and the correlation ρ. For fixed p the call option prices C^i solve the parabolic differential equation

$$\frac{s^2 y}{2}\frac{\partial^2 C^i}{\partial s^2} + \rho s \xi y \frac{\partial^2 C^i}{\partial s \partial y} + \frac{\xi^2 y}{2}\frac{\partial^2 C^i}{\partial y^2} + rs\frac{\partial C^i}{\partial s} + \kappa(\theta - y)\frac{\partial C^i}{\partial y} + \frac{\partial C^i}{\partial t} = rC^i, \quad (7.1)$$
$$(t, s, y) \in (0, T_i) \times (0, \infty) \times (0, \infty)$$

with initial and boundary conditions

$$\begin{aligned} C^i(T_i, s, y) &= (s - K_i)^+ \ \ \forall\, (s, y) \\ C^i(t, 0, y) = 0,\ \frac{\partial C^i}{\partial s}&(t, \infty, y) = 1 \ \ \forall\, (t, y) \\ rs\frac{\partial C^i}{\partial s}(t, s, 0) + \kappa\theta\frac{\partial C^i}{\partial y}(t, s, 0) + \frac{\partial C^i}{\partial t}(t, s, 0) &= rC^i(t, s, 0) \ \ \forall\, (t, s) \\ C^i(t, s, \infty) &= s \ \ \forall\, (t, s). \end{aligned}$$

As mentioned before, the feasible set of the linear semi-infinite optimization problem above is closed, convex and due to the box constraints also compact. Furthermore the feasible set is nonempty if the standard call with maturity $T_i = T$ and strike $K_i = K$ is included as call C^j in the hedge portfolio, because then the strategy $\alpha_j = 1, \alpha_i = 0$, $i \ne j$ satisfies all constraints. This in turn implies that a solution of the optimization problem exists.

In the following we refer to the robust static super-replication problem in its compact form

$$\begin{aligned} &\min_{\alpha \in \mathbb{R}^{n+1}} \ c^{\mathrm{T}}\alpha \\ \text{s.t.}\quad &a_1(t, y, p)^{\mathrm{T}}\alpha \ge 0 \quad \forall\, (t, y, p) \in [0, T] \times [Y_{\min}, Y_{\max}] \times U \\ &a_2(s)^{\mathrm{T}}\alpha \ge b_2(s) \quad \forall\, s \in [0, D] \\ &\alpha_i^{\mathrm{lb}} \le \alpha_i \le \alpha_i^{\mathrm{ub}}, \quad i = 0, \dots, n \end{aligned} \qquad \text{(P)}$$

with suitable vector- and scalar-valued functions a_1, a_2 and b_2. In analogy to the arguments in Chapter 5 we observe that the second constraint $a_2(s)^{\mathrm{T}}\alpha \ge b_2(s)$ is model-independent and hence can be neglected in the context of a robustification. In contrast to this the performance of the optimal hedge portfolio crucially depends on the model prices $C^i(t, D, y, p)$ included in the vector a_1. If the uncertainty set $[Y_{\min}, Y_{\max}] \times U$ contains the implied model parameters (y, p) at the time of the barrier hit, the model prices will at least be close to the market prices. However, small deviations in the size of a few basis points are likely and hence should ideally be considered during the optimization.

Thus we now robustify the solution of optimization problem (P) with respect to perturbations of the vector a_1 in the sense of model errors by asking the corresponding inequality to hold in a small ellipsoid around the model prices. These ellipsoids shall be defined by associated matrices $E(t, y, p)$. This results in the following robust optimization problem.

Definition 7.1. Consider problem (P) of finding the cost-optimal superhedge with suitable vectors c, a_1, a_2 and scalars b_2. For $(t, y, p) \in [0, T] \times [Y_{\min}, Y_{\max}] \times U$ let $E(t, y, p) \in \mathbb{R}^{(n+1)\times(n+1)}$ be positive definite matrices defining ellipsoids around the model prices $a_1(t, y, p)$. Further assume that the map $E : [0, T] \times [Y_{\min}, Y_{\max}] \times U \to \mathbb{R}^{(n+1)\times(n+1)}$ is continuous. Then the robust counterpart of problem (P) is given by the optimization problem

$$
\begin{aligned}
&\min_{\alpha \in \mathbb{R}^{n+1}} \ c^{\mathrm{T}}\alpha \\
\text{s.t.}\quad &(a_1(t, y, p) + E(t, y, p)u)^{\mathrm{T}}\alpha \geq 0 \ \ \forall\, u \in \mathbb{R}^{n+1} : \|u\|_2 \leq 1 \\
&\qquad \forall\, (t, y, p) \in [0, T] \times [Y_{\min}, Y_{\max}] \times U =: \Gamma_1 \\
&\qquad a_2(s)^{\mathrm{T}}\alpha \geq b_2(s) \ \ \forall\, s \in [0, D] \\
&\qquad \alpha_i^{\mathrm{lb}} \leq \alpha_i \leq \alpha_i^{\mathrm{ub}}, \ \ i = 0, \ldots, n.
\end{aligned}
\tag{RP}
$$

Note that optimization problem (RP) introduces the robustness against model errors in the price space, whereas the model parameter uncertainty was eliminated in the parameter space of the model. This mathematically shows the different quality and treatment of these two separate sources of risk. First we make use of the financial market model to generate likely volatility surface scenarios for varying model parameters. In the next step we put small ellipsoids around the model option prices which geometrically leads to all volatility surface combinations that are close to one of the model volatility surfaces.

In the simplest case the matrix $E(t, y, p)$ defining the ellipsoids might be chosen as a small multiple of the identity matrix, but in a more realistic setting the model error and hence the matrix will depend on time t and the parameters (y, p). In the following we assume that the matrices $E(t, y, p)$ are chosen in such a way that the feasible set of problem (RP) is nonempty. Due to the compactness of the feasible set we then readily obtain that a solution of the optimization problem exists.

Assumption 7.2. Assume that the feasible set of problem (RP) is nonempty.

Note that problem (RP) is still a linear semi-infinite optimization problem, but now the complexity of the problem has increased significantly due to the additional variable

u varying in the $(n+1)$-dimensional unit ball. Hence it would be very desirable to reduce the dimension of the semi-infinite parameter set by eliminating u from problem (RP). The next theorem shows that this can actually be achieved.

Theorem 7.3. *The linear semi-infinite optimization problem (RP) is equivalent to the semi-infinite second order cone problem*

$$\begin{aligned} &\min_{\alpha \in \mathbb{R}^{n+1}} \ c^{\mathrm{T}}\alpha \\ \text{s.t.} \quad & a_1(t,y,p)^{\mathrm{T}}\alpha \geq \|E(t,y,p)^{\mathrm{T}}\alpha\|_2 \ \ \forall\, (t,y,p) \in \Gamma_1 \\ & a_2(s)^{\mathrm{T}}\alpha \geq b_2(s) \ \ \forall\, s \in [0,D] \\ & \alpha_i^{\mathrm{lb}} \leq \alpha_i \leq \alpha_i^{\mathrm{ub}}, \ \ i = 0,\ldots,n. \end{aligned} \tag{RP–SOCP}$$

Proof. Similar to Ben-Tal and Nemirovski [9], we prove the equivalence by showing that for every $(t,y,p) \in \Gamma_1$ the infinite number of constraints

$$(a_1(t,y,p) + E(t,y,p)u)^{\mathrm{T}}\alpha \geq 0 \ \ \forall\, u \in \mathbb{R}^{n+1} : \|u\|_2 \leq 1 \tag{7.2}$$

is equivalent to the single constraint $a_1(t,y,p)^{\mathrm{T}}\alpha \geq \|E(t,y,p)^{\mathrm{T}}\alpha\|$. Note that (7.2) can be rewritten as

$$\begin{aligned} &\min_{u \in \mathbb{R}^{n+1}:\|u\|_2 \leq 1} \ (a_1(t,y,p) + E(t,y,p)u)^{\mathrm{T}}\alpha \\ &\qquad = \min\nolimits_{u \in \mathbb{R}^{n+1}:\|u\|_2 \leq 1} \left(a_1(t,y,p)^{\mathrm{T}}\alpha + (E(t,y,p)^{\mathrm{T}}\alpha)^{\mathrm{T}}u\right) \geq 0. \end{aligned}$$

It is easy to prove that the minimum of this linear function on the unit circle is attained for the vector $u_* = -E(t,y,p)^{\mathrm{T}}\alpha / \|E(t,y,p)^{\mathrm{T}}\alpha\|_2$ if $E(t,y,p)^{\mathrm{T}}\alpha \neq 0$. However, as $E(t,y,p)$ is a positive definite matrix, the case $E(t,y,p)^{\mathrm{T}}\alpha = 0$ would imply $\alpha = 0$ which is not admissible for problem (RP), because then the second constraint in this problem would imply $a_2(s)^{\mathrm{T}}\alpha = a_2(s)^{\mathrm{T}}0 = 0 \geq b_2(s) = (s-K)^+ > 0$ for $s \in (K,D]$. Hence (7.2) is equivalent to

$$\begin{aligned} &(a_1(t,y,p) + E(t,y,p)u_*)^{\mathrm{T}}\alpha \\ &\qquad = a_1(t,y,p)^{\mathrm{T}}\alpha - (E(t,y,p)^{\mathrm{T}}\alpha)^{\mathrm{T}} \frac{E(t,y,p)^{\mathrm{T}}\alpha}{\|E(t,y,p)^{\mathrm{T}}\alpha\|_2} \geq 0. \end{aligned}$$

This expression in turn can be transformed to $a_1(t,y,p)^{\mathrm{T}}\alpha \geq \|E(t,y,p)^{\mathrm{T}}\alpha\|_2$ which proves the theorem. □

The new equivalent formulation (RP–SOCP) of (RP) also allows to interpret the robustness added to (P). If the first constraint of (RP–SOCP) is compared to the corresponding constraint of (P), it is clear that the hedge portfolio described by the robust problem offers a safety margin $\|E(t,y,p)^{\mathrm{T}}\alpha\|_2 > 0$ that protects the portfolio against model errors.

Further note that problem (RP–SOCP) is still a semi-infinite optimization problem, but compared to problem (RP) the parameter set $\{u \in \mathbb{R}^{n+1} : \|u\|_2 \leq 1\} \times \Gamma_1$ has been reduced to Γ_1. The price for this reduction is the nonlinearity and non-smoothness which now enters the constraints of problem (RP–SOCP) in the form of second order cone constraints.

In Section 7.2 we will present an alternative problem formulation which reduces the complexity of the parameter space while at the same time preserving the linear structure of the underlying problem (RP). But before we turn to the associated transformations, we will briefly discuss the numerical solution of problems (RP) and (RP–SOCP). As the two optimization problems are equivalent, we will focus on the reduced problem (RP–SOCP).

In general an algorithm solving problem (RP–SOCP) will replace the infinite number of constraints associated with the parameter sets $[0,T]\times[Y_{\min}, Y_{\max}]\times U$ and $[0,D]$ by a discrete set of constraints. Let the discrete approximations of these parameter sets be denoted by $M_1 \subset [0,T] \times [Y_{\min}, Y_{\max}] \times U$ and $M_2 \subset [0,D]$, $|M_1|, |M_2| < \infty$. The resulting discrete version of (RP–SOCP) is then given by

$$\begin{aligned} &\min_{\alpha\in\mathbb{R}^{n+1}} \; c^{\mathrm{T}}\alpha \\ \text{s.t.} \;\; & a_1(t,y,p)^{\mathrm{T}}\alpha \geq \|E(t,y,p)^{\mathrm{T}}\alpha\|_2 \;\; \forall\, (t,y,p) \in M_1 \\ & a_2(s)^{\mathrm{T}}\alpha \geq b_2(s) \;\; \forall\, s \in M_2 \\ & \alpha_i^{\mathrm{lb}} \leq \alpha_i \leq \alpha_i^{\mathrm{ub}}, \;\; i = 0,\dots,n. \end{aligned} \qquad \text{(RP–SOCP–DISCR)}$$

Clearly, problem (RP–SOCP–DISCR) is a second order cone program where we can assure the existence of solutions for any discrete sets M_1, M_2 due to the compactness of the associated feasible set and Assumption 7.2. Hence problem (RP–SOCP–DISCR) can be solved by standard SOCP- or SDP-solvers. However, the solution of this problem is in general not feasible for the original problem (RP–SOCP). To reduce the infeasibility, we adapt the exchange method of semi-infinite programming to our particular problem and successively add the most violating constraints to the sets M_1, M_2. This leads to Algorithm 4.

To compute a solution of problem (RP–SOCP), the algorithm successively solves second order cone programs (RP–SOCP–DISCR) and nonlinear optimization problems in the form of (CV–SOCP). Note that the latter problem is in fact a PDE-constrained optimization problem, because the vector a_1 depends on the price functions $C^i(t,D,y,p)$ which in turn are the solution of the partial differential equation (7.1). The next theorem shows that these subproblems and hence Algorithm 4 is well defined and that each limit point of the sequence $(\alpha^k)_k$ is an optimal solution of problem (RP–SOCP).

Theorem 7.4. *Assume that Assumption 7.2 holds. Then Algorithm 4 is well defined. Furthermore, if* $\epsilon = 0$, *every limit point of the sequence* $(\alpha^k)_{k\in\mathbb{N}}$ *is an optimal solution of problem (RP–SOCP).*

Algorithm 4: SOCP-NLP Cutting Plane Discretization

Input: Let $M_1 \subset [0,T] \times [Y_{\min}, Y_{\max}] \times U$ and $M_2 \subset [0,D]$, $|M_1|, |M_2| < \infty$ be given initial grids. Further let $\epsilon > 0$ be a suitable convergence tolerance and $k = 0$.

Main Algorithm:

(S1) Calculate an optimal solution α^k of the discretized problem (RP–SOCP–DISCR).

(S2) Determine the constraint violation (CV) of α^k for problem (RP–SOCP) by minimizing the slack-functions at α^k:

$$\delta_1 = \min_{(t,y,p)\in\Gamma_1} a_1(t,y,p)^{\mathrm{T}}\alpha^k - \|E(t,y,p)^{\mathrm{T}}\alpha^k\|_2 \qquad \text{(CV–SOCP)}$$

$$\delta_2 = \min_{s\in[0,D]} a_2(s)^{\mathrm{T}}\alpha^k - b_2(s)$$

If $\min\{\delta_1, \delta_2\} \geq -\epsilon$ then STOP.

(S3) Add the minimizers of the slack functions (the most violating constraints) to M_1, M_2. Set $k \leftarrow k+1$ and go to step (S1).

Proof. As mentioned before, step (S1) is well defined due to Assumption 7.2. Furthermore, the two slack minimizations in step (S2) always have a solution, because in these problems continuous functions are minimized on compact sets. The map $E : [0,T] \times [Y_{\min}, Y_{\max}] \times U \to \mathbb{R}^{(n+1)\times(n+1)}$ was assumed to be continuous and the continuity of a_1, a_2 can be verified in analogy to the proof of Lemma 5.3.

Note that, due to Theorem 7.3, optimization problem (RP–SOCP) is equivalent to the linear semi-infinite optimization problem (RP). Hence it is sufficient to show, that Algorithm 4 in fact produces iterates of an algorithm for the solution of a linear semi-infinite optimization problem and then apply the corresponding convergence theory.

The proof of Theorem 7.3 shows, that the solution of the discretized problem (RP–SOCP–DISCR) in step (S1) of Algorithm 4 is equivalent to the linear semi-infinite programming problem

$$\begin{aligned} &\min_{\alpha\in\mathbb{R}^{n+1}} c^{\mathrm{T}}\alpha \\ \text{s.t.}\quad &(a_1(t,y,p) + E(t,y,p)u)^{\mathrm{T}}\alpha \geq 0 \;\; \forall\, (u,t,y,p) \in \{u : \|u\|_2 \leq 1\} \times M_1 \\ &a_2(s)^{\mathrm{T}}\alpha \geq b_2(s) \;\; \forall\, s \in M_2 \\ &\alpha_i^{\mathrm{lb}} \leq \alpha_i \leq \alpha_i^{\mathrm{ub}},\;\; i = 0,\ldots,n. \end{aligned} \tag{7.3}$$

Furthermore, due to Theorem 7.3 and its proof, the problem of computing the worst case constraint violation (CV–SOCP) in step (S2) is equivalent to

$$\min_{(u,t,y,p)\in\{u:\|u\|_2\leq 1\}\times\Gamma_1} (a_1(t,y,p)+E(t,y,p)u)^{\mathrm{T}}\alpha, \tag{7.4}$$

because one can eliminate the variable u from the minimization by explicitly computing the optimal $u_*(t,y,p) := -E(t,y,p)^{\mathrm{T}}\alpha/\|E(t,y,p)^{\mathrm{T}}\alpha\|_2$. In particular, if (t_*,y_*,p_*) denotes the optimal solution of problem (CV–SOCP), then the candidate $(u_*(t_*,y_*,p_*),t_*,y_*,p_*)$ is also the solution of (7.4) and vice versa.

Adding the point (t_*,y_*,p_*) to the set M_1 in step (S3) of Algorithm 4 and hence the constraint $a_1(t_*,y_*,p_*)^{\mathrm{T}}\alpha \geq \|E(t_*,y_*,p_*)^{\mathrm{T}}\alpha\|_2$ to the next SOCP in step (S1), is equivalent to adding the infinite number of constraints

$$(a_1(t_*,y_*,p_*)+E(t_*,y_*,p_*)u)^{\mathrm{T}}\alpha \geq 0 \quad \forall\, u \in \{u : \|u\|_2 \leq 1\}$$

to problem (7.3). Hence the vectors (u,t_*,y_*,p_*), $u \in \{u : \|u\|_2 \leq 1\}$, can be interpreted as an infinite number of feasibility cuts for the linear semi-infinite optimization problem which are added to the parameter sets $\{u : \|u\|_2 \leq 1\} \times M_1$.

Applying the standard convergence theory of linear semi-infinite optimization (see e.g. Theorem 11.2 in Goberna and Lopez [46]) to a successive solution of problems (7.3), (7.4) and with the mesh update $\{u : \|u\|_2 \leq 1\}\times(M_1\cup\{t_*,y_*,p_*\})$ mentioned previously then proves the convergence. □

The proof of Theorem 7.4 made heavy use of the equivalence of the linear semi-infinite optimization problem (RP) and the semi-infinite second order cone problem (RP–SOCP). For the numerical implementation, it seems to be advantageous to solve problem (RP–SOCP) instead of (RP), because the dimension of the semi-infinite parameter space $\{u \in \mathbb{R}^{n+1} : \|u\|_2 \leq 1\}\times[0,T]\times[Y_{\min},Y_{\max}]\times U$ is reduced drastically from typically $15-30$ (depending on the number of financial products n included in the portfolio) to 6, the dimension of $[0,T]\times[Y_{\min},Y_{\max}]\times U$. Correspondingly, the global optimization problems (CV–SOCP) can be carried out in a much smaller space which can be expected to significantly reduce computation time. However, from a computational point of view the drawback is the additional nonlinearity in the form of the second order cone constraints. In the next section we will see how the additional SOCP-nonlinearity can be eliminated from the optimization problems arising in Algorithm 4.

7.2 Equivalent SDP–NSDP Formulation

In this section we restate the robust second order cone problem (RP–SOCP) as a semi-infinite semidefinite programming problem. As it turns out, this equivalent formulation can be used to derive an iterative procedure similar to Algorithm 4 which successively solves SDPs instead of the second order cone problems (RP–SOCP–DISCR) and NS-DPs instead of the nonlinear programming problems (CV–SOCP). Based on Theorem 7.4 we can prove convergence for this mixed SDP–NSDP procedure.

It is a well-known fact that for fixed $(t, y, p) \in [0, T] \times [Y_{\min}, Y_{\max}] \times U$ a conic quadratic constraint of the form $a_1(t, y, p)^{\mathrm{T}}\alpha \ge \|E(t, y, p)^{\mathrm{T}}\alpha\|_2$ can be explicitly converted to an SDP constraint. The following result states the equivalent SDP formulation of the second order cone problem (RP–SOCP). For completeness we present a detailed proof.

Lemma 7.5. *The conic quadratic problem (RP–SOCP) is equivalent to the following optimization problem with an infinite number of SDP constraints*

$$\begin{aligned} &\min_{\alpha \in \mathbb{R}^{n+1}} \; c^{\mathrm{T}}\alpha \\ \text{s.t.} \quad &\mathcal{A}(t, y, p; \alpha) \succeq 0 \;\; \forall \, (t, y, p) \in [0, T] \times [Y_{\min}, Y_{\max}] \times U \\ &a_2(s)^{\mathrm{T}}\alpha \ge b_2(s) \quad \forall \, s \in [0, D] \\ &\alpha_i^{\mathrm{lb}} \le \alpha_i \le \alpha_i^{\mathrm{ub}}, \quad i = 1, \ldots, n \end{aligned} \tag{RP–SDP}$$

with the matrix $\mathcal{A}(t, y, p; \alpha)$ defined by

$$\mathcal{A}(t, y, p; \alpha) := \begin{pmatrix} a_1(t, y, p)^{\mathrm{T}}\alpha \, I_{n+1} & -E(t, y, p)^{\mathrm{T}}\alpha \\ -\alpha^{\mathrm{T}}E(t, y, p) & a_1(t, y, p)^{\mathrm{T}}\alpha \end{pmatrix} \in \mathbb{S}^{n+2}, \tag{7.5}$$

where $I_{n+1} \in \mathbb{R}^{(n+1)\times(n+1)}$ denotes the identity matrix and $\mathbb{S}^{n+2}$ is the space of all real symmetric $(n + 2) \times (n + 2)$ matrices.

Proof. To show the equivalence of (RP–SOCP) and (RP–SDP) it is sufficient to prove that

$$\mathcal{A}(t, y, p; \alpha) \succeq 0 \quad \Longleftrightarrow \quad a_1(t, y, p)^{\mathrm{T}}\alpha \ge \|E(t, y, p)^{\mathrm{T}}\alpha\|_2.$$

Suppose that $\mathcal{A}(t, y, p; \alpha) \succeq 0$. Then we obtain for all vectors $z = (\xi, \tau)^{\mathrm{T}} \in \mathbb{R}^{n+2}$ ($\xi \in \mathbb{R}^{n+1}, \tau \in \mathbb{R}$)

$$0 \le z^{\mathrm{T}}\mathcal{A}(t, y, p; \alpha)z = a_1(t, y, p)^{\mathrm{T}}\alpha||\xi||_2^2 - 2\tau\alpha^{\mathrm{T}}E(t, y, p)\xi + a_1(t, y, p)^{\mathrm{T}}\alpha\tau^2. \tag{7.6}$$

Assuming $a_1(t, y, p)^{\mathrm{T}}\alpha < 0$, then for $\xi \in \mathbb{R}^{n+1}\backslash\{0\}$ and $\tau = 0$ the relation (7.6) would imply $0 \le z^{\mathrm{T}}\mathcal{A}(t, y, p; \alpha)z = a_1(t, y, p)^{\mathrm{T}}\alpha \; ||\xi||_2^2 < 0$ which is a contradiction. Hence $\mathcal{A}(t, y, p; \alpha) \succeq 0$ always yields $a_1(t, y, p)^{\mathrm{T}}\alpha \ge 0$.

If $a_1(t, y, p)^{\mathrm{T}}\alpha = 0$, we can deduce from (7.6) for $\xi = E(t, y, p)^{\mathrm{T}}\alpha$ and $\tau > 0$ that $E(t, y, p)^{\mathrm{T}}\alpha = 0$, since in this case inequality (7.6) yields $0 \le z^{\mathrm{T}}\mathcal{A}(t, y, p; \alpha)z = -2\tau||E(t, y, p)^{\mathrm{T}}\alpha||_2^2$ which in turn is equivalent to $\|E(t, y, p)^{\mathrm{T}}\alpha\|_2 = 0$ ($\Longleftrightarrow \alpha = 0$ since $E(t, y, p) \succ 0$). Therefore, we get $a_1(t, y, p)^{\mathrm{T}}\alpha \ge 0 = \|E(t, y, p)^{\mathrm{T}}\alpha\|_2$ in the case of $a_1(t, y, p)^{\mathrm{T}}\alpha = 0$.

Otherwise, if $a_1(t, y, p)^{\mathrm{T}}\alpha > 0$, then choosing $z = (E(t, y, p)^{\mathrm{T}}\alpha, a_1(t, y, p)^{\mathrm{T}}\alpha)^{\mathrm{T}}$ in (7.6) yields

$$0 \le z^{\mathrm{T}}\mathcal{A}(t, y, p; \alpha)z = a_1(t, y, p)^{\mathrm{T}}\alpha \left(\left(a_1(t, y, p)^{\mathrm{T}}\alpha\right)^2 - \|E(t, y, p)^{\mathrm{T}}\alpha\|_2^2\right),$$

which also implies $a_1(t,y,p)^{\mathrm{T}}\alpha \geq \|E(t,y,p)^{\mathrm{T}}\alpha\|_2$.

On the other hand, suppose $a_1(t,y,p)^{\mathrm{T}}\alpha \geq \|E(t,y,p)^{\mathrm{T}}\alpha\|_2$, then for every $z = (\xi,\tau)^{\mathrm{T}} \in \mathbb{R}^{n+2}$ we deduce by using the Cauchy–Schwarz formula

$$-\tau\left(E(t,y,p)^{\mathrm{T}}\alpha\right)^{\mathrm{T}}\xi \geq -|\tau|\left|\langle E(t,y,p)^{\mathrm{T}}\alpha,\xi\rangle_2\right| \geq -|\tau|\|E(t,y,p)^{\mathrm{T}}\alpha\|_2\,\|\xi\|_2,$$

which in turn implies

$$\begin{aligned}
z^{\mathrm{T}}\mathcal{A}(t,y,p;\alpha)z &= a_1(t,y,p)^{\mathrm{T}}\alpha\ \|\xi\|_2^2 - 2\tau\,\alpha^{\mathrm{T}}E(t,y,p)\xi + a_1(t,y,p)^{\mathrm{T}}\alpha\ \tau^2\\
&\geq\ a_1(t,y,p)^{\mathrm{T}}\alpha\ \|\xi\|_2^2 - 2|\tau|\underbrace{\|E(t,y,p)^{\mathrm{T}}\alpha\|_2}_{\leq a_1(t,y,p)^{\mathrm{T}}\alpha}\ \|\xi\|_2 + a_1(t,y,p)^{\mathrm{T}}\alpha\ \tau^2\\
&\geq\ a_1(t,y,p)^{\mathrm{T}}\alpha\left(\tau^2 - 2|\tau|\,\|\xi\|_2 + \|\xi\|_2^2\right)\ \geq\ a_1(t,y,p)^{\mathrm{T}}\alpha\left(|\tau| - \|\xi\|_2\right)^2\ \geq\ 0.
\end{aligned}$$

Thus $a_1(t,y,p)^{\mathrm{T}}\alpha \geq \|E(t,y,p)^{\mathrm{T}}\alpha\|_2$ implies $\mathcal{A}(t,y,p;\alpha) \succeq 0$. □

Based on Lemma 7.5 the idea arises to exploit the equivalent semidefinite matrix constraint for the numerical solution of the optimization problems (RP–SOCP) and (RP–SDP). In analogy to Algorithm 4 we can hence discretize the parameter space of the constraints in problem (RP–SDP) and solve the resulting semidefinite programming problem. But then the question remains if the equivalent matrix constraint can also be used to compute the worst case constraint violation in step (S2) of the algorithm. Intuitively, to compute the slack of problem (RP–SDP) we need to solve the following eigenvalue problem:

$$\min_{(t,y,p)\in[0,T]\times[Y_{\min},Y_{\max}]\times U} \lambda_{\min}\left(\mathcal{A}(t,y,p;\alpha)\right), \tag{CV–SDP}$$

where $\alpha \in \mathbb{R}^{n+1}$ is assumed to be given, $\mathcal{A}(t,y,p;\alpha) \in \mathbb{S}^{n+2}$ is defined by (7.5) and $\lambda_{\min}\left(\mathcal{A}(t,y,p;\alpha)\right)$ denotes the minimal eigenvalue of $\mathcal{A}(t,y,p;\alpha)$. Using the next lemma we can show the equivalence of (CV–SOCP) and the eigenvalue problem (CV–SDP).

Lemma 7.6. *Let $I_{n+1} \in \mathbb{R}^{(n+1)\times(n+1)}$ be the identity matrix, $d \in \mathbb{R}^{n+1}$, $\beta \in \mathbb{R}$ and the matrix A be given by*

$$A := \begin{pmatrix} \beta I_{n+1} & d \\ d^{\mathrm{T}} & \beta \end{pmatrix} \in \mathbb{S}^{n+2}.$$

Then the eigenvalues $\lambda_i(A)$, $i = 1,\ldots,n+2$ of A are $\lambda_{\min}(A) := \lambda_1(A) = \beta - \|d\|_2$, $\lambda_2(A) = \ldots = \lambda_{n+1}(A) = \beta$, and $\lambda_{\max}(A) := \lambda_{n+2}(A) = \beta + \|d\|_2$.

Proof. First we assume that $\lambda \neq \beta$ ($\lambda \in \mathbb{R}$) and determine the roots of the characteristic polynomial of $A - \lambda I_{n+2}$, that is

$$\det\left(A - \lambda I_{n+2}\right) = \det\begin{pmatrix} (\beta-\lambda)I_{n+1} & d \\ d^{\mathrm{T}} & \beta-\lambda \end{pmatrix} = 0.$$

Now we apply a Gaussian elimination step to the last row of $A - \lambda I_{n+2}$. Defining $0 = (0, \dots, 0)^{\mathrm{T}} \in \mathbb{R}^{n+1}$ we obtain

$$\begin{aligned} \det(A - \lambda I_{n+2}) &= \det\begin{pmatrix} (\beta-\lambda) I_{n+1} & d \\ 0^{\mathrm{T}} & (\beta-\lambda) - \sum_{i=1}^{n+1} \frac{d_i^2}{\beta-\lambda} \end{pmatrix} \\ &= (\beta-\lambda)^{n+1} \left((\beta-\lambda) - \sum_{i=1}^{n+1} \frac{d_i^2}{\beta-\lambda} \right) = 0, \end{aligned}$$

which in turn is equivalent to

$$0 = (\beta-\lambda)^n \left((\beta-\lambda)^2 - \|d\|^2 \right) \quad \Longleftrightarrow \quad (\beta-\lambda)^2 = \|d\|^2,$$

in case of $\lambda \neq \beta$. Thus $\lambda_1(A) = \beta - \|d\|_2$, $\lambda_{n+2}(A) = \beta + \|d\|_2$ are two eigenvalues of A. If $\lambda = \beta$, then we deduce by applying the Laplacian expansion theorem

$$\begin{aligned} \det\begin{pmatrix} (\beta-\lambda) I_{n+1} & d \\ d^{\mathrm{T}} & \beta-\lambda \end{pmatrix} &= \det\begin{pmatrix} 0_{(n+1)\times(n+1)} & d \\ d^{\mathrm{T}} & 0 \end{pmatrix} = \\ &= \sum_{j=1}^{n+1} (-1)^{n+2+j} d_j \det\left(A_{j,n+2}\right) = 0, \end{aligned}$$

where $A_{j,n+2} \in \mathbb{R}^{(n+1)\times(n+1)}$ is a submatrix of A obtained by deleting row j and column $(n+2)$ from A. Thus, β is an eigenvalue of A and it is straightforward to show that the multiplicity of this eigenvalue is equal to n. Hence we have $\lambda_2(A) = \dots = \lambda_{n+1}(A) = \beta$. Ordering the eigenvalues $\lambda_1, \dots, \lambda_{n+2}$ yields the desired result. □

Based on these findings the following lemma proves that the objective function values of (CV–SOCP) and (CV–SDP) coincide.

Lemma 7.7. *Let $\mathcal{A}(t, y, p; \alpha) \in \mathbb{S}^{n+2}$ be defined by (7.5), then*

i) $\lambda_{\min}(\mathcal{A}(t, y, p; \alpha)) = a_1(t, y, p)^{\mathrm{T}} \alpha - \|E(t, y, p)^{\mathrm{T}} \alpha\|_2$

ii) (CV–SOCP) $\Longleftrightarrow$ (CV–SDP).

Proof. The result immediately follows by applying Lemma 7.6 to $\mathcal{A}(t, y, p; \alpha)$. □

Due to the previous lemma, the nonlinear minimization problem (CV–SOCP) is equivalent to the minimization of the minimal eigenvalue of $\mathcal{A}(t, y, p; \alpha)$, where this matrix depends in a very nonlinear fashion on the parameters $(t, y, p) \in [0, T] \times [Y_{\min}, Y_{\max}] \times U$. In general such an eigenvalue minimization problem is a very hard non-smooth optimization problem (see e.g. [21, 22]).

In some cases it is well known that eigenvalue optimization problems can be transformed to an SDP. For example the minimization of the maximal eigenvalue of a linear matrix function can be restated as a semidefinite program (see e.g. [86]). However, in our case we are interested in minimizing the *minimal* eigenvalue of a nonlinear matrix function, which contrasts the typical goal in other research areas like robust control design (see e.g. [22,49,62,63,78]).

In the next theorem we show, by using duality arguments, that the problem of minimizing the minimal eigenvalue of a matrix function can also be rewritten as a matrix optimization problem with SDP constraints. To the knowledge of the author, no general result for this case is known so far.

Theorem 7.8. *For $k, l, m \in \mathbb{N}$ let $\mathcal{M} : X \to \mathbb{S}^m$, $X \subseteq \mathbb{R}^{k \times l}$, be a real symmetric matrix function. Then the following two problems are equivalent:*

$$\min\nolimits_{x \in X} \ \lambda_{\min}\left(\mathcal{M}(x)\right), \tag{7.7}$$

$$\min\nolimits_{(x,\Sigma) \in X \times \mathbb{S}^m} \left\{ \mathrm{Tr}(\Sigma \ \mathcal{M}(x)) \ \mid \ \Sigma \succeq 0, \ 1 - \mathrm{Tr}(\Sigma) = 0 \right\}, \tag{7.8}$$

where $\lambda_{\min}(M)$ denotes the minimal eigenvalue of $M \in \mathbb{S}^m$ and $\mathrm{Tr}(M)$ is the trace operator applied to $M \in \mathbb{R}^{m \times m}$, respectively. Moreover, the optimal function values coincide.

Proof. We prove the theorem in two steps. First we show that the computation of the minimal eigenvalue of a matrix function is equivalent to the maximization of a scalar variable subject to a linear matrix inequality constraint. In a second step we derive the dual of this maximization problem which completes the proof.

First, fixing $x \in X$ we show

$$\lambda_{\min}\left(\mathcal{M}(x)\right) = \max_{\tau \in \mathbb{R}} \left\{ \tau \ \mid \ \mathcal{M}(x) - \tau I_m \succeq 0 \right\}, \tag{7.9}$$

where $I_m \in \mathbb{R}^{m \times m}$ is the identity matrix. The Raleigh–Ritz theorem (see for example [53, Theorem 4.2.2]) implies $M - \lambda_{\min} \ I_m \succeq 0$, where $\lambda_{\min} := \lambda_{\min}\left(\mathcal{M}(x)\right)$ and $M := \mathcal{M}(x)$. To show that $\lambda_{\min}$ is the sharpest bound satisfying the matrix inequality $M - \tau \ I_m \succeq 0$ we prove the following equivalence

$$M \succeq \tau \ I_m \quad \Longleftrightarrow \quad \lambda_{\min}(M) \geq \tau, \tag{7.10}$$

where $M \in \mathbb{S}^m$ and $\tau \in \mathbb{R}$. To show this we diagonalize the real symmetric matrix $M \in \mathbb{S}^m$. Let $H \in \mathbb{R}^{m \times m}$ be an orthogonal matrix such that

$$H^{\mathrm{T}} M H = \mathrm{diag}(\lambda_1, \ldots, \lambda_m) =: D,$$

where $\lambda_{\min}(M) := \lambda_1 \leq \lambda_2 \leq \ldots \leq \lambda_m$ are the real eigenvalues of M and $\mathrm{diag}(\lambda_1, \ldots, \lambda_m) \in \mathbb{R}^{m \times m}$ denotes a diagonal matrix with diagonal components $\lambda_1, \ldots, \lambda_m$ (see e.g. [53, Corollary 2.5.14]). Using this decomposition we get

$$M - \tau \ I_m = HDH^{\mathrm{T}} - \tau \ I_m = H(D - \tau \ I_m)H^{\mathrm{T}} \succeq 0 \quad \Longleftrightarrow \quad D - \tau \ I_m \succeq 0,$$

which in turn is equivalent to

$$\lambda_i \geq \tau \quad \forall\, i = 1, \ldots, m \qquad \Longleftrightarrow \qquad \lambda_{\min}(M) \geq \tau.$$

This proves (7.10) and therefore (7.9).

Secondly, using (7.9), we prove the equivalence of (7.7) and (7.8) by the strong Wolfe duality theorem (see, e.g. [39, Theorem 9.5.1], [87]). The Lagrangian of the primal problem (7.9) is defined by $\ell : \mathbb{R} \times \mathbb{S}^m \to \mathbb{R}$,

$$\ell(\tau, \Sigma) = \tau + \langle \Sigma, \mathcal{M}(x) - \tau I_m \rangle = (1 - \mathrm{Tr}(\Sigma))\ \tau + \mathrm{Tr}(\Sigma\ \mathcal{M}(x)).$$

The first order Fréchet derivative of ℓ with respect to τ applied to $\delta\tau$ is given by

$$\ell'_\tau(\tau, \Sigma)\delta\tau = \langle \delta\tau, \nabla_\tau \ell(\tau, \Sigma) \rangle = (1 - \mathrm{Tr}(\Sigma))\, \delta\tau$$

and thus we obtain the following (necessary and sufficient) optimality conditions for problem (7.9)

$$\nabla_\tau \ell(\tau, \Sigma) = 1 - \mathrm{Tr}(\Sigma) = 0,\ \Sigma \succeq 0,\ \mathrm{Tr}\,(\Sigma(\mathcal{M}(x) - \tau I_m)) = 0,\ \mathcal{M}(x) - \tau I_m \succeq 0.$$

Applying the strong duality theorem we further obtain the dual of problem (7.9)

$$\min_{(\tau,\Sigma)\in\mathbb{R}\times\mathbb{S}^m} \quad \ell(\tau, \Sigma) \quad \text{s.t.}\ \ 1 - \mathrm{Tr}(\Sigma) = 0, \quad \Sigma \succeq 0.$$

Using $1 - \mathrm{Tr}(\Sigma) = 0$, the objective function of this dual problem reduces to

$$\ell(\tau, \Sigma) = (1 - \mathrm{Tr}(\Sigma))\ \tau + \mathrm{Tr}(\Sigma\ \mathcal{M}(x)) = \mathrm{Tr}(\Sigma\ \mathcal{M}(x)).$$

Therefore, the final version of the dual problem is given by the following SDP

$$\min_{\Sigma\in\mathbb{S}^m}\ \mathrm{Tr}(\Sigma\ \mathcal{M}(x)) \quad \text{s.t.}\ \ 1 - \mathrm{Tr}(\Sigma) = 0, \quad \Sigma \succeq 0. \tag{7.11}$$

Moreover, the strong duality result ensures that the objective function values of the primal (linear) SDP (7.9) and the dual (linear) SDP (7.11) are equal (for fixed, but arbitrary $x \in X$), that is

$$\begin{aligned}\lambda_{\min}(\mathcal{M}(x)) &= \max_{\tau\in\mathbb{R}}\ \{\tau \mid \mathcal{M}(x) - \tau I_m \succeq 0\} = \\ &= \min_{\Sigma\in\mathbb{S}^m}\ \{\mathrm{Tr}(\Sigma\ \mathcal{M}(x)) \mid 1 - \mathrm{Tr}(\Sigma) = 0,\ \Sigma \succeq 0\}.\end{aligned}$$

But this relation implies the desired result, i.e. now we know that

$$\begin{aligned}\min_{x\in X}\ \{\lambda_{\min}(\mathcal{M}(x))\} &= \min_{x\in X}\left\{\max_{\tau\in\mathbb{R}}\ \{\tau \mid \mathcal{M}(x) - \tau I_m \succeq 0\}\right\} \\ &= \min_{x\in X}\left\{\min_{\Sigma\in\mathbb{S}^m}\ \{\mathrm{Tr}(\Sigma\ \mathcal{M}(x)) \mid 1 - \mathrm{Tr}(\Sigma) = 0,\ \Sigma \succeq 0\}\right\} \\ &= \min_{(x,\Sigma)\in X\times\mathbb{S}^m}\ \{\mathrm{Tr}(\Sigma\ \mathcal{M}(x)) \mid 1 - \mathrm{Tr}(\Sigma) = 0,\ \Sigma \succeq 0\}.\end{aligned}$$

Hence, problem (7.7) is equivalent to (7.8). □

Now we apply the previous general theorem to the particular case of determining the worst case constraint violation for (RP–SDP). In particular we prove that, due to the nonlinearity of the matrix function $\mathcal{A}(\cdot,\cdot,\cdot;\alpha)$ (α fixed), the eigenvalue problem (CV–SDP) is equivalent to a *nonlinear* semidefinite programming problem.

Corollary 7.9. *The problem of computing the constraint violation (CV–SOCP) is equivalent to the nonlinear semidefinite program*

$$\begin{aligned} \min_{(t,y,p,\Sigma)\in\Gamma_1\times\mathbb{S}^{n+2}} \quad & \left(a_1(t,y,p)^{\mathrm{T}}\alpha\right)\, \mathrm{Tr}(\Sigma) - 2\alpha^{\mathrm{T}}E(t,y,p)d_\Sigma \\ \textit{s.t.} \quad & \Sigma \succeq 0, \quad \mathrm{Tr}(\Sigma)=1, \end{aligned} \tag{CV–NSDP}$$

where, for $\Sigma_1 \in \mathbb{S}^{n+1}$, $d_\Sigma \in \mathbb{R}^{n+1}$ and $\beta_\Sigma \in \mathbb{R}$, the matrix $\Sigma \in \mathbb{S}^{n+2}$ is defined by

$$\Sigma := \begin{pmatrix} \Sigma_1 & d_\Sigma \\ d_\Sigma^{\mathrm{T}} & \beta_\Sigma \end{pmatrix}.$$

Proof. In Lemma 7.7 we showed that

$$\text{(CV–SOCP)} \iff \text{(CV–SDP)}.$$

Applying Theorem 7.8 to (CV–SDP), we obtain the equivalence of problem (CV–SDP) and the NSDP

$$\begin{aligned} \min_{(t,y,p,\Sigma)\in[0,T]\times[Y_{\min},Y_{\max}]\times U\times\mathbb{S}^{n+2}} \quad & \mathrm{Tr}\left(\Sigma\ \mathcal{A}(t,y,p;\alpha)\right) \\ \text{s.t.} \quad & \Sigma\succeq 0, \quad 1-\mathrm{Tr}(\Sigma)=0. \end{aligned} \tag{7.12}$$

It remains to verify that the objective of this NSDP and (CV–NSDP) coincide. Due to the special structure of $\mathcal{A}(t,y,p;\alpha)\in\mathbb{S}^{n+2}$, we set

$$\mathcal{A}(t,y,p;\alpha) = \begin{pmatrix} a_1(t,y,p)^{\mathrm{T}}\alpha\ I_{n+1} & -E(t,y,p)^{\mathrm{T}}\alpha \\ -\alpha^{\mathrm{T}}E(t,y,p) & a_1(t,y,p)^{\mathrm{T}}\alpha \end{pmatrix} =: \begin{pmatrix} \beta I_{n+1} & d \\ d^{\mathrm{T}} & \beta \end{pmatrix}$$

and compute

$$\begin{aligned} \Sigma\mathcal{A}(t,y,p;\alpha) &= \begin{pmatrix} \Sigma_1 & d_\Sigma \\ d_\Sigma^{\mathrm{T}} & \beta_\Sigma \end{pmatrix}\begin{pmatrix} \beta I_{n+1} & d \\ d^{\mathrm{T}} & \beta \end{pmatrix} \\ &= \begin{pmatrix} \beta\Sigma_1 + d_\Sigma d^{\mathrm{T}} & \Sigma_1 d + \beta d_\Sigma \\ \beta d_\Sigma^{\mathrm{T}} + \beta_\Sigma d^{\mathrm{T}} & d_\Sigma^{\mathrm{T}} d + \beta_\Sigma\beta \end{pmatrix}. \end{aligned}$$

Finally $\beta_\Sigma = \Sigma_{n+1,n+1}$ and $\mathrm{Tr}(d_\Sigma d^{\mathrm{T}}) = d_\Sigma^{\mathrm{T}} d$ implies

$$\begin{aligned} \mathrm{Tr}\left(\Sigma\ \mathcal{A}(t,y,p;\alpha)\right) &= \mathrm{Tr}\left(\beta\Sigma_1 + d_\Sigma d^{\mathrm{T}}\right) + d_\Sigma^{\mathrm{T}} d + \beta_\Sigma\beta \\ &= \beta\ \mathrm{Tr}\left(\Sigma_1\right) + \beta\beta_\Sigma + \mathrm{Tr}\left(d_\Sigma d^{\mathrm{T}}\right) + d_\Sigma^{\mathrm{T}} d \\ &= \beta\ \mathrm{Tr}\left(\Sigma\right) + 2d_\Sigma^{\mathrm{T}} d \end{aligned}$$

by using properties of the trace operator. Hence, we can rewrite the objective function of (7.12) to

$$\mathrm{Tr}\,(\Sigma\ \mathcal{A}(t,y,p;\alpha)) = \left(a_1(t,y,p)^{\mathrm{T}}\alpha\right)\ \mathrm{Tr}(\Sigma) - 2\alpha^{\mathrm{T}}E(t,y,p)d_{\Sigma},$$

which proves the corollary. □

Combining all the previous results we can restate Algorithm 4 in an equivalent SDP–NSDP formulation. While Algorithm 4 successively solves a sequence of second order cone programs (SOCP) and nonlinear programming problems (NLP), the restated Algorithm 5 iteratively solves a sequence of SDPs and NSDPs.

Algorithm 5: SDP-NSDP Cutting Plane Discretization

Input: Let $M_1 \subset [0,T] \times [Y_{\min}, Y_{\max}] \times U$ and $M_2 \subset [0,D]$, $|M_1|, |M_2| < \infty$ be given initial grids. Further let $\epsilon > 0$ be a suitable convergence tolerance and $k = 0$.

Main Algorithm:

(S1) Calculate an optimal solution α^k of the (linear) semidefinite program

$$\begin{aligned} &\min_{\alpha\in\mathbb{R}^{n+1}}\ c^{\mathrm{T}}\alpha \\ \text{s.t.}\ \ &\mathcal{A}(t,y,p;\alpha) \succeq 0\ \ \forall\ (t,y,p) \in M_1 \\ &a_2(s)^{\mathrm{T}}\alpha \geq b_2(s)\ \ \forall\ s \in M_2 \\ &\alpha_i^{\mathrm{lb}} \leq \alpha_i \leq \alpha_i^{\mathrm{ub}},\ \ i = 0,\dots,n \end{aligned} \qquad \text{(RP–SDP–DISCR)}$$

(S2) Determine the constraint violation of α^k for problem (RP–SDP) by minimizing the slack-functions at α^k:

$$\begin{aligned} \delta_1 = &\min_{t,y,p,\Sigma} (a_1(t,y,p)^{\mathrm{T}}\alpha^k)\mathrm{Tr}(\Sigma) - 2(\alpha^k)^{\mathrm{T}}E(t,y,p)d_{\Sigma} \qquad \text{(CV–NSDP)} \\ &\text{s.t.}\ (t,y,p,\Sigma) \in \Gamma_1 \times \mathcal{S}^{n+2},\ \Sigma \succeq 0,\ \mathrm{Tr}(\Sigma) = 1 \\ \delta_2 = &\min_{s\in[0,D]} a_2(s)^{\mathrm{T}}\alpha^k - b_2(s) \end{aligned}$$

If $\min\{\delta_1, \delta_2\} \geq -\epsilon$ then STOP.

(S3) Add the minimizers of the slack functions (the most violating constraints) to M_1, M_2. Set $k \leftarrow k+1$ and go to step (S1).

Comparing Algorithms 4 and 5, the nonlinear SOCP-constraint in problem (RP–SOCP–DISCR) is replaced by a linear matrix inequality in problem (RP–SDP–DISCR). Furthermore the SOCP-nonlinearity is also removed from problem (CV–SOCP) for the price of an additional matrix variable in the objective function and simple linear matrix constraints in problem (CV–NSDP). In particular the objective functions of the constraint violation problems look very similar. Furthermore, by applying Theorem 7.4, we immediately obtain the following convergence theorem for Algorithm 5.

Corollary 7.10. *Assume that Assumption 7.2 holds. Then Algorithm 5 is well defined. Furthermore, if* $\epsilon = 0$*, every limit point of the sequence* $(\alpha^k)_{k\in\mathbb{N}}$ *is an optimal solution of problem (RP–SDP).*

Proof. We prove the corollary by showing that each step of Algorithm 5 is equivalent to the corresponding step in Algorithm 4. For step (S1), the equivalence of (RP–SDP–DISCR) and (RP–SOCP–DISCR) follows directly by the proof of Lemma 7.5. The equivalence of steps (S2) was proven in Corollary 7.9. Therefore, we deduce the convergence of the algorithm from Theorem 7.4. □

7.3 Numerical Results

The previous sections showed that static hedging strategies with robustness against model errors can theoretically be computed by solving a sequence of second order cone and nonlinear optimization problems or alternatively a sequence of semidefinite and nonlinear semidefinite programs. However, so far it is unclear whether these algorithms can also successfully be implemented in practice. In this section we will show that it is in fact possible to solve optimization problem (RP) by making use of standard optimization software. Moreover, we will analyze and interpret the resulting static hedging strategies from a practitioner's point of view.

To better compare our results with existing hedging strategies we pick up the example of Subsection 5.2.3 of hedging an up-and-out call with strike $K = 2750$, barrier $D = 3300$ and a maturity of $T = 1$ year. The underlying of the barrier option is the EURO STOXX 50 index with price $S_0 = 2750$ in September 2004. The Heston model parameters are chosen as follows: The risk-free interest rate is defined to be $r = 5.5\%$, the dividend yield $\delta = 2.5\%$, the start variance $Y_0 = 0.04$, the long run mean of the variance $\theta_0 = 0.04$, the mean reversion speed $\kappa_0 = 1.5$, the volatility of volatility $\xi_0 = 0.2$ and the correlation $\rho_0 = -0.5$. For this set of parameters, the fair value of the considered up-and-out call is 1.60% in percent of the underlying S_0.

The parameters mentioned above specify the value $C_i(0, S_0, Y_0, \kappa_0, \theta_0, \xi_0, \rho_0)$ of the calls in the portfolio at time $t = 0$ and hence also the vector c defining the objective function in (RP). However, the Heston model parameters $(y, p) = (y, \kappa, \theta, \xi, \rho)$ at the time of a barrier hit can be expected to differ from the initial parameters $(Y_0, p_0) := (Y_0, \kappa_0, \theta_0, \xi_0, \rho_0)$. These differences are caused by changes of the shape of the volatility surface which can constantly be observed in the market. As in Subsection 5.2.3, we therefore ask the super-replication property to hold for a whole set $[Y_{\min}, Y_{\max}] \times U$ of model parameters given as the intersection of the hypercube $[Y_{\min}, Y_{\max}] \times [\kappa_{\min}, \kappa_{\max}] \times$

$[\theta_{\min}, \theta_{\max}] \times [\xi_{\min}, \xi_{\max}] \times [\rho_{\min}, \rho_{\max}]$ centered around (Y_0, p_0) and the set of points satisfying the Feller condition $\kappa\theta - \xi^2/2 \geq 0$. In the following we set the diameter of the individual one-dimensional intervals uniformly to $\Delta := \sqrt{Y_{\max}} - \sqrt{Y_{\min}} = \kappa_{\max} - \kappa_{\min} = \sqrt{\theta_{\max}} - \sqrt{\theta_{\min}} = \xi_{\max} - \xi_{\min} = \rho_{\max} - \rho_{\min} = 15\%$. This problem data, together with the simple bounds $\alpha_i^{\text{lb}} = -50$ and $\alpha_i^{\text{ub}} = 50$ fully specifies optimization problem (P), whose solution was computed in Subsection 5.2.3 (see Table 5.16) to be given by the optimal portfolio weights listed in Table 7.1.

	C_1	C_2	C_3	C_4	C_5	C_6	C_7	C_8	C_9	C_{10}
T_i	1.00	1.00	1.00	1.00	1.00	0.75	0.75	0.75	0.50	0.50
K_i	2750	3300	3350	3450	3600	3300	3400	3600	3300	3500
α_i	1.00	−30.42	37.95	−9.13	0.47	0.04	0.31	−0.29	0.01	0.10

Table 7.1: Optimal Portfolio Weights α_i solving Problem (P)

The computed portfolio consists of (only) ten standard calls which are liquidly traded on the EUREX. Furthermore, the portfolio is robust against movements of the implied Heston parameters within a 15% range. Although this robustness guarantees the hedge performance for a large variety of shapes of the volatility surface, the super-replication portfolio is surprisingly cheap (1.84%). In particular it is only 24 basis points more expensive than the Heston price of the up-and-out call.

However, although these properties are very desirable, the super-replication property of the portfolio listed in Table 7.1 is limited to future call option prices which exactly match a model volatility surface scenario corresponding to some model parameter $(y, p) \in [Y_{\min}, Y_{\max}] \times U$. Consequently, the super-replication property may be lost if the market prices at the time of a barrier hit deviate from the call prices of the calibrated Heston model at that time. Since we cannot expect the Heston model to perfectly fit market data, small differences of model and market prices are likely. To robustify the hedge portfolio against these price deviations, we now consider the solution of problem (RP).

In order to fully specify problem (RP) we have to define the matrices $E(t, y, p)$ describing the ellipsoidal uncertainty sets around the Heston model prices $a_1(t, y, p)$. Two specifications of this matrix immediately come to one's mind. A first idea is to uniformly robustify each call option price in the sense of absolute price deviations. This can be achieved by choosing $E(t, y, p) = E_{\text{abs}}(t, y, p) := \mu I_n$, $\mu > 0$, as a small multiple of the identity matrix. The disadvantage of this specification is that it does not take the magnitude of the call option prices into account. For example, if the barrier in the above example is hit shortly before maturity (that means the value of the underlying is equal to $D = 3300$), the price of the call with a maturity of one year and strike 3600 is close to zero, because the probability of the underlying rising above 3600 in the remaining short time period is negligible. Consequently, a robustification against absolute price deviations may be too conservative for options with large strikes and and too mild for options with small strikes. This motivates to robustify the hedge against relative price deviations, which corresponds to the matrix $E(t, y, p) = E_{\text{rel}}(t, y, p) := \mu \cdot \text{diag}(a_1(t, y, p))$, $\mu > 0$. In the following we will present results for both specifications $E(t, y, p) = E_{\text{abs}}(t, y, p)$ and $E(t, y, p) = E_{\text{rel}}(t, y, p)$.

Note that the bounds $\alpha_i^{\text{lb}} = -50$ and $\alpha_i^{\text{ub}} = 50$ guarantee that the trading strategy $\alpha_1 = 1, \alpha_i = 0, i \neq 1$, solely consisting of the call C_1 with the same strike and maturity as the barrier option is feasible for problem (RP) if $\mu > 0$ in the definition of the matrices E_{abs}, E_{rel} is not too large. This can best be observed by looking at problems (RP), (P) and noting that $C_1(t, D, y, p)$ is uniformly greater than some positive constant for $(t, y, p) \in [0, T] \times [Y_{\min}, Y_{\max}] \times U$. Thus Assumption 7.2 is fulfilled such that a solution of optimization problem (RP) exists. Furthermore, Theorem 7.4 and Corollary 7.10 imply that every limit point of the sequence $(\alpha^k)_{k \in \mathbb{N}}$ generated by Algorithms 4 and 5 is an optimal solution of problem (RP).

Due to the lack of generally applicable NSDP solvers we will in the following tackle problem (RP) with the SOCP-based Algorithm 4. In our implementation the solutions of the second order cone programs (RP–SOCP–DISCR) in step (S1) of the algorithm are computed with the interior-point solver SDPT3 [84] which we conveniently called from Matlab with the cvx modeling system for convex optimization [48]. The nonlinear optimization problems (CV–SOCP) were solved with Matlab's standard fmincon solver contained in the optimization toolbox. Table 7.2 shows the resulting iteration output of Algorithm 4 for the example of a robustification with $E(t, y, p) = E_{\text{abs}}(t, y, p)$ and $\mu = 0.02\%$.

Iteration	Slack	Cost(α^k)	$\lvert M_1\rvert + \lvert M_2\rvert$
0	−3.387791e−001	7.758802e−003	865
1	−3.818223e−002	2.056644e−002	867
2	−4.524836e−003	2.119374e−002	872
3	−7.412633e−004	2.133473e−002	876
4	−7.138447e−005	2.144145e−002	879
5	−8.718422e−005	2.144295e−002	883
6	−1.707001e−005	2.144324e−002	886
7	−3.814816e−006	2.144334e−002	889
8	−9.019763e−007	2.144339e−002	896
9	−2.198364e−007	2.144341e−002	898
10	−6.161411e−008	2.144342e−002	900

Table 7.2: Sample Iteration Process of Algorithm 4 for $E(t, y, p) = E_{\text{abs}}(t, y, p)$ and $\mu = 0.02\%$

Table 7.2 shows that Algorithm 4 successfully terminates after 10 iterations with a slack satisfying the convergence tolerance $\epsilon = 10^{-7}$. The algorithm starts with initial grids M_1, M_2 containing 865 nodes and successively adds the most violated constraints to these meshes by solving the nonlinear optimization problems (CV–SOCP). Note that the number of nodes in the mesh only slightly increases during the iteration process. As more and more constraints are added, the portfolio α^k approaches feasibility until the optimal solution (see Table 7.3) with a cost of 2.14% is reached.

	C^1	C^2	C^3	C^4	C^5	C^6	C^7	C^8	C^9	C^{10}
T_i	1.00	1.00	1.00	1.00	1.00	0.75	0.75	0.75	0.50	0.50
K_i	2750	3300	3350	3450	3600	3300	3400	3600	3300	3500
α_i	1.00	−16.67	12.29	7.15	−4.25	0.38	−0.51	0.65	0.02	−0.03

Table 7.3: Optimal Portfolio Weights α_i solving problem (RP) with $E(t, y, p) = E_{\text{abs}}(t, y, p)$ and $\mu = 0.02\%$

We note that although the number of second order cone constraints approaches one thousand in the example above, the main effort of Algorithm 4 is not the solution of the SOCPs, but instead the computation of the worst case constraint violation. Problem (CV–SOCP) is computationally demanding since it is a global optimization problem which we solve by a repeated solution of nonlinear programs with random start iterate. In summary the solution of the price-robust problem (RP) can be computed in approximately five minutes on a usual PC, which seems surprisingly fast regarding the complexity of the underlying optimization problem with an infinite number of second order cone constraints.

Apart from these optimization insights, the optimal solution listed in Table 7.3 also reveals some properties which are important from a financial point of view. First of all the robustification in the price space increases the cost of the (parameter-robust) portfolio by approximately 30 basis points. This premium assures the super-replication property of the portfolio if the individual model prices differ at most two basis points from the corresponding market prices. The resulting additional robustness is also graphically visible in Figure 7.1 which illustrates the value of the portfolios listed in Tables 7.1 and 7.3 for various spot prices S at maturity $T = 1$ (note that we restrict the plot to those options with maturity of one year since all other options have previously expired worthless). Compared to the solution of problem (P) the price-robust portfolio makes use of a much less aggressive call spread to prevent losses that might result from price deviations in the model and market prices of the calls C^2 and C^3.

Intuitively it is clear that an increased robustness (a larger parameter μ) will also increase the price of the portfolio. To analyze this effect in more detail, Table 7.4 illustrates the cost of the optimal portfolios α^* solving the robust optimization problem (RP) with varying degree of price-robustness.

Table 7.4 shows that the cost of portfolios with absolute price robustness grows linearly in the robustness parameter μ. In contrast to this the relative robustness seems to be less costly for large μ. This makes sense from a financial point of view, because price deviations of cheap calls (with high strikes) are over-emphasized by a robustification against absolute price changes. This in particular applies for the case of a barrier hit shortly before maturity of the barrier option, in which case the value of options with strikes above the barrier is close to zero. The treatment of relative price

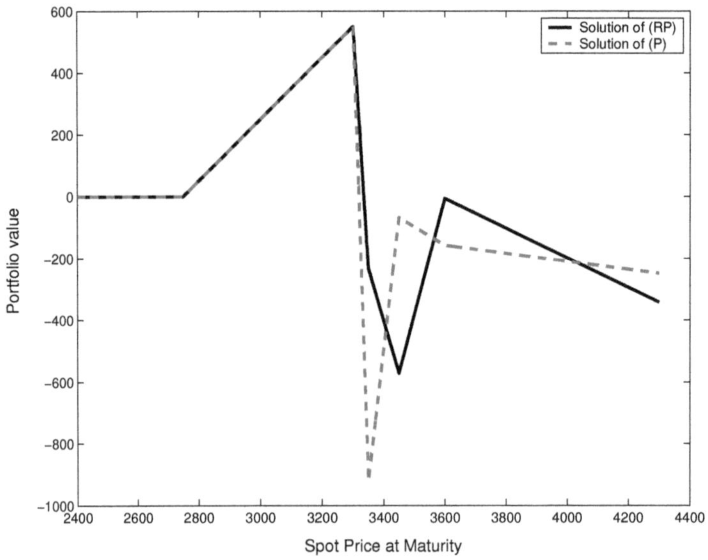

Figure 7.1: Graphical illustration of the strike-spread component of the portfolios listed in Tables 7.1 and 7.3

$E_{\text{abs}}(t, y, p)$	μ	0.005%	0.01%	0.02%	0.03%	0.04%
	$\text{cost}(\alpha^*)$	1.93%	2.02%	2.14%	2.22%	2.27%
$E_{\text{rel}}(t, y, p)$	μ	0.1%	0.2%	0.3%	0.4%	0.5%
	$\text{cost}(\alpha^*)$	1.94%	2.04%	2.11%	2.15%	2.19%

Table 7.4: Cost of portfolios solving problem (RP) with varying absolute and relative price-robustness.

deviations leads to less aggressive uncertainty sets for these calls which in turn results in cheaper hedge portfolios.

From a practical point of view, a robustification against price deviations in the magnitude of a few basis points is sufficient, because calibration errors for a portfolio consisting of ten standard options are usually very small. Hence, for the absolute price robustness $\mu = 0.01\%$ or $\mu = 0.02\%$ can be expected to suffice, whereas the relative price robustness only requires to choose $\mu = 0.1\%$ or $\mu = 0.2\%$ (we note that the relative price robustness implied by $\mu = 0.2\%$ corresponds to an absolute robustness in the size of several basis points for option C^1).

To summarize, the static super-replication problem (RP) for an up-and-out call with additional robustness against model errors can be solved by a successive SDP–NSDP or SOCP–NLP algorithm. The convergence of the method was obtained by showing that the standard theory of linear semi-infinite optimization can be applied to the equivalent problem formulations.

The presented algorithms solve a sequence of discretized versions of the underlying semi-infinite SDP or SOCP and nonlinear minimization problems for the computation of the worst case constraint violation of the iterates. As it turned out, the latter problem can be reformulated in the SDP case as the minimization of the minimal eigenvalue of a nonlinear matrix function. This eigenvalue problem was proven to be equivalent to a nonlinear semidefinite program, where the nonlinearity is introduced by the parameters of a parabolic differential equation. As a by-product we obtained a reformulation of the minimization of the minimal eigenvalue of a general matrix function as a semidefinite matrix optimization problem with linear matrix constraints.

Numerical results showed that the proposed algorithm can successfully be applied to robustify static hedge portfolios for barrier options against model errors. The computed portfolios reduce the weights of the instruments used for the call spread and thus provide a safety cushion protecting the bank issuing the barrier option against potential deviations of model and market prices.

Although we presented the solution procedure for a particular financial market application, the proposed method can be applied in analogy to other optimization problems with an infinite number of linear constraints and ellipsoidal uncertainty sets around the problem parameters. As a key finding of this chapter, nonlinear semidefinite programs naturally arise in the computation of the constraint violation of robust optimization problems which motivates to develop general NSDP-solvers. Since these are currently not available we were forced to implement the SOCP–NLP based algorithm. However, it would be interesting to compare this implementation to the equivalent SDP–NSDP-based algorithm. In particular further research is needed to investigate how much the solution of the constraint violation problem (which is a global optimization problem) benefits from an elimination of the SOCP nonlinearity.

8 Empirical Hedge Performance

In the previous chapters we presented various methods to robustify static hedge portfolios against changes of the implied volatility surface. Numerical results underlined that the computed static hedge portfolios are surprisingly cheap compared to the embedded degree of robustness against model parameter uncertainty or model errors. However, so far we did not demonstrate how the hedge portfolio performs in practice. This chapter will close this gap by presenting the results of an empirical analysis carried out by Maruhn, Nalholm and Fengler in [68].

To be more precise the empirical study consists of monitoring the hedging error of the robust static superhedge for up-and-out calls and down-and-out puts (written on the DAX index) with one year maturity and initiated at every day of a seven year dataset. For each of these barrier options we monitor the hedge performance if either the barrier is breached or the option expires.

Since the market prices of the hedge instruments constantly change throughout the sample, one would expect that it is necessary to recompute the portfolio weights for static hedges of barrier options issued on different days. However, we are able to show that the robust static super-replication method allows to derive minimal assumption hedges which are not only robust against changes of the volatility surface, but also independent of the hedge instrument prices at initiation. This enables us to set up the same static hedge (in the sense of portfolio weights) for all barrier options issued throughout the sample.

Our empirical analysis shows that the minimal assumption hedge is truly robust against changes of the volatility surface. If we assume that the hedge portfolio is liquidated on the barrier, the super-replication property holds in any market environment covered by the seven year dataset. For the case of up-and-out calls the hedge is super-replicating even in the presence of moves beyond the barrier in the sample. For down-and-out puts the larger magnitude of downward moves leads to rare but extreme hedge errors which shows that this kind of barrier options has to be robustified against skew and gap risk. To underline the performance of the robust static hedge, we compare our results to the strike spread approach of Carr and Chou [23] and a dynamic delta-vega hedge based on the local volatility model. Surprisingly, the minimal assumption hedge provides the smallest dispersion at an acceptable additional cost.

The chapter is structured as follows. In Section 8.1 we present the large and current dataset of volatility surfaces on the DAX used for the empirical study. Afterwards Section 8.2 describes how we set up the robust super-replicating hedge and its minimal assumption version for the analysis. Other hedging strategies that we use for comparisons are described in Section 8.3, followed by Section 8.4 which introduces the methodology of the empirical investigation. Finally, Section 8.5 presents the results along with an evaluation of the different hedging approaches.

8.1 Description of the Dataset

The dataset used for the empirical study consists of daily implied volatility data for vanilla options on the DAX traded at the EUREX. The sample period reaches from January 2000 to November 2006 covering a total of 1752 observations. The options have European exercise and are some of the most liquid index contracts in Europe. Since the DAX is a capital weighted performance index, dividends can be set to zero when valuing options on the DAX, see Deutsche Börse [14].

The observed implied volatility corresponds to the last traded price on a given day. To prevent stale prices the exchange establishes an official settlement price if the last traded price is older than 15 minutes or if for some reason it is deemed not to reflect prevailing market conditions. To further clean the data, implied volatilities of options with less than 10 days to expiry are removed from the sample as are implied volatilities exceeding 80%. Our dynamic hedging benchmark will make use of a local volatility model and it is therefore vital that there are no arbitrage opportunities in the cleaned dataset. To prevent this, the data is smoothed using the procedure in Fengler [37]. After the smoothing, a regular grid of implied volatilities is constructed. For interest rates, the daily zero coupon rates are linearly interpolated as is commonly done, see e.g. Dumas et al. [28].

The level of the DAX and the 1-year at-the-money (ATM) implied volatilities are shown in Figure 8.1. The sample exhibits periods with a trending market, periods with high as well as low volatility and periods with extreme movements like the 8.89% drop on September 11, 2001. The level of both the short and long (1 year) end of the yield curve varies between 2% and 5% in the sample period.

When considering reverse barrier options, it is not only the implied volatility level that is of interest. For example an up-and-out call (UOC) can be decomposed into a vanilla call capped at the barrier and a position in a one-touch option struck at the barrier. The latter is sensitive to the ATM skew of the implied volatility surface at a touch event. In Figure 8.2, we plot the term structures of the average, minimum, maximum and 2.5% as well as 97.5% quantiles of the ATM skew of the implied volatility surface. Figure 8.2 also shows the time series of daily observations for the 1-week and 1-year ATM skews in the implied volatility.

The stable flattish skew term structure at the long end as well as the volatile steeply sloped short end of the skew term structure are consistent with stylized facts of option markets on equity indices as treated in e.g. Gatheral [43]. From Figure 8.2 it may be seen that models lacking considerable variation in the short term skew fail to capture a very real phenomenon in the market. Popular equity models such as the Heston stochastic volatility model [50] as well as the jump-diffusion stochastic volatility model widely used e.g. by Bates [6], Bakshi et al. [5], Eraker et al. [35] and Eraker [34] are too rigid to capture this variation in the skew. This is particularly problematic for hedges of reverse barrier options, where the area near the barrier and close to expiry is highly exposed to the short end of the skew term structure.

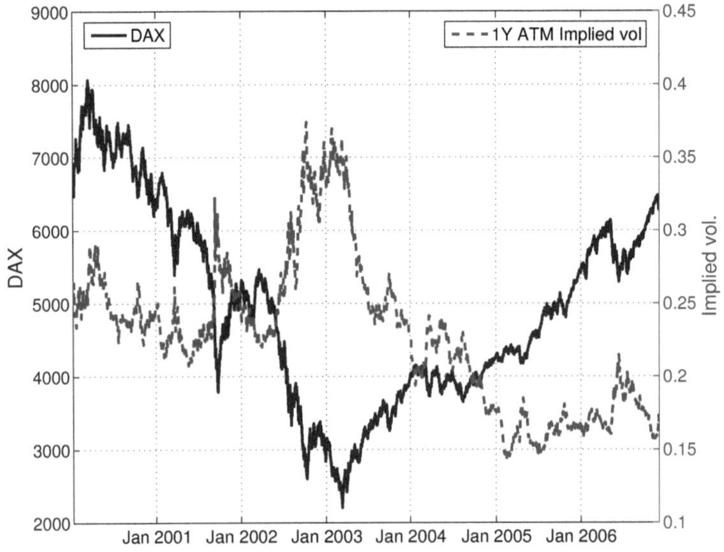

Figure 8.1: DAX level and 1-Year At-the-Money implied volatility.

The shortcoming of typical models to capture features such as the variation in the implied volatility skew has led to the near-universal practice of recalibration. This approach tries to (partly) offset the limited model dynamics by refitting the model to prices of liquidly traded instruments after changes of market data, and hence in particular to include more realistic volatility surface dynamics into the computation of the Greeks.

The calibration of models such as the Heston stochastic volatility (HSV) model described by the stochastic differential equations

$$
\begin{aligned}
dS_t &= (r-\delta)S_t dt + \sqrt{Y_t} S_t dW_t^1 \\
dY_t &= \kappa(\theta - Y_t)dt + \xi\sqrt{Y_t} dW_t^2, \quad dW_t^1 dW_t^2 = \rho dt \\
dB_t &= rB_t dt
\end{aligned}
\tag{8.1}
$$

is an inverse problem and as such requires some regularization, see Gerlich, Giese, Maruhn and Sachs [44] or Bouchouev and Isakov [15] for a review of the local volatility case.

In Figure 8.3 we plot time series of parameters for the Heston Stochastic Volatility model calibrated to the daily DAX implied volatility surfaces. For the calibration we used vega-weighted implied volatilities and fixed $\kappa = 2$ as proposed by Bergomi [11]. The parameter tracks are qualitatively in accordance with the results of Buehler [20] (see Figure 5.7) as well as Bergomi [11] who calibrate to the EURO STOXX 50 index.

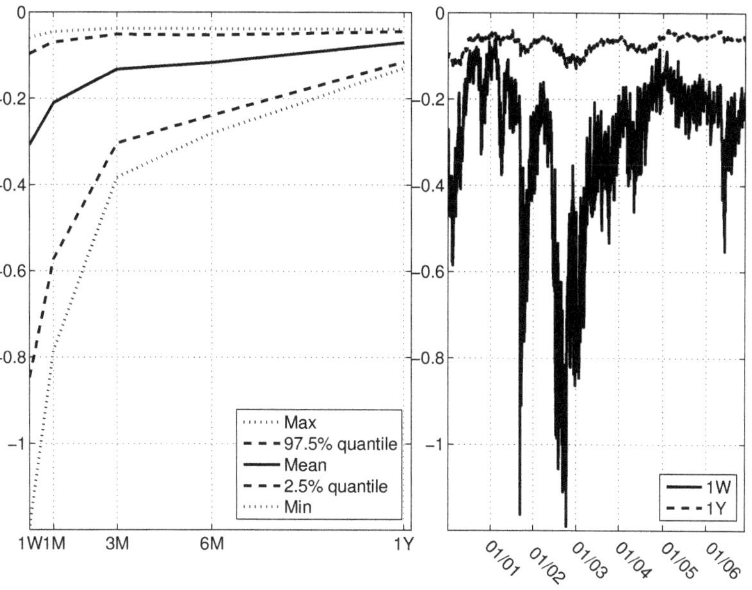

Figure 8.2: Average term structures and full time series of At-The-Money implied volatility skew on the DAX.

It is worth noting the agreement between the relative changes of the one week ATM volatility skew in Figure 8.2 and those in the time series for ρ in Figure 8.3. The role of ρ in the HSV model is to capture the skew, so the agreement is to be expected. Even though the volatility of variance ξ also has (to a certain degree) impact on the skew, ρ is still the dominating parameter. Similarly, note the agreement between the one year ATM volatility in Figure 8.1 and Y_0 in Figure 8.3. Again, this agreement is to be expected since Y_0, together with θ captures the volatility level. Overall, the instable parameter tracks simply reflect the fluctuations of the market and underline the necessity to recalibrate the model in order to fit it to market data.

However, there is an inherent inconsistency of the recalibration practice. If the calibrated parameters change then, essentially, this is equivalent to adopting a new model at each recalibration. From Figure 8.3, it is clear that hedges based on a recalibrated HSV model are likely to face substantial deterioration of the hedge performance even if one tries to hedge model parameter changes in the sense of sensitivities.

To make a hedge strategy robust to such parameter changes is of major importance to avoid hedge errors caused by changing market conditions. One interesting approach is that of Avellaneda, Levy and Parás [4] where the Black–Scholes model is essentially made robust to uncertainty about the key volatility parameter. However, since static hedge portfolios consist of options with several strikes and expiries, a robust hedge

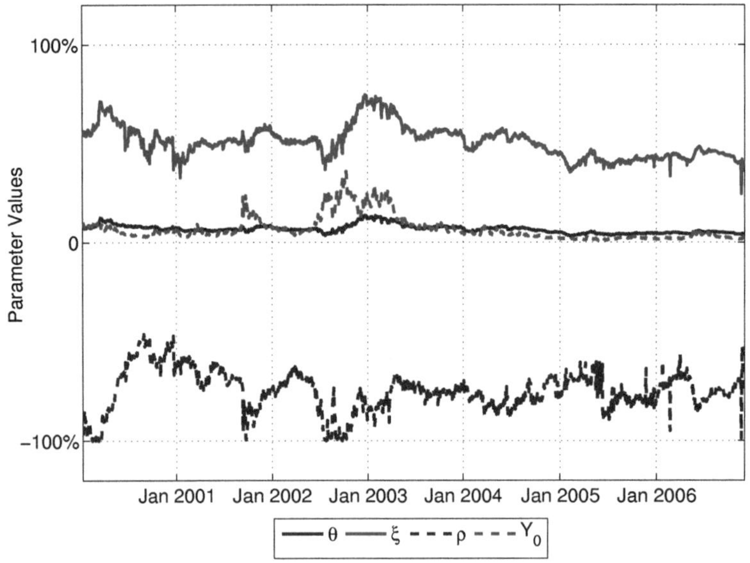

Figure 8.3: Time series of parameters in the Heston model when calibrated to daily DAX volatility surfaces. Speed of mean-reversion was fixed at $\kappa = 2$ in the calibration.

cannot ignore the overall shape of the implied volatility surface and only focus on the (current and future) instantaneous volatility as in the uncertain volatility approach. The necessity to also take variations of skew and smile parameters into account underlines the importance of the robust static super-replication approach presented in Section 5.2. An analysis of the impact of the skew on the static super-replication portfolio in the next section will lead to further insights into the skew sensitivity of barrier options and their static hedges.

8.2 Setting up the Robust Static Hedge

In this section we describe the robust static super-replication portfolio used within the scope of the empirical analysis. In Subsection 8.2.1 we examine the effects and cost of a successive robustification and thus motivate the degree of robustness we incorporate into the static hedge portfolio. Afterwards Subsection 8.2.2 introduces a variant of the hedge which allows us to use the same portfolio weights for all barrier options initiated within the seven year dataset.

8.2.1 Step-Wise Robustification

The observed real-world dynamics in the previous section raise the question how the weights α_i and the cost of the super-replication hedge defined in (5.9) change if we robustify the portfolio against different features of the implied volatility surface. In the following this will be investigated by performing the robustification in three steps as indicated in Table 8.1. The first step (A) basically robustifies against changes in the level of the implied volatility surface in both the short and long term. The hedge is made robust to the three parameters $\sqrt{Y_0}$, κ, $\sqrt{\theta}$ in one step since the portfolio is quite stable at each incremental step. The second step (B) additionally robustifies against changes in the smile of the implied volatility surface. Finally, the third step (C) adds robustness against changes in the implied volatility skew.

<table>
<tr><th colspan="2">Parameter</th><th>p_0</th><th>Lower bound</th><th>Upper bound</th><th colspan="3">Step</th></tr>
<tr><td>1</td><td>$\sqrt{Y_0}$</td><td>0.25</td><td>0.10</td><td>0.40</td><td rowspan="3">A</td><td rowspan="4">B</td><td rowspan="5">C</td></tr>
<tr><td>2</td><td>κ</td><td>2.00</td><td>1.50</td><td>2.50</td></tr>
<tr><td>3</td><td>$\sqrt{\theta}$</td><td>0.26</td><td>0.20</td><td>0.30</td></tr>
<tr><td>4</td><td>ξ</td><td>0.50</td><td>0.20</td><td>0.60</td></tr>
<tr><td>5</td><td>ρ</td><td>−0.70</td><td>−0.90</td><td>−0.40</td></tr>
</table>

Table 8.1: Heston parameter intervals used for the robustification.

The initial parameter values $(\sqrt{Y_0}, p_0) = (\sqrt{Y_0}, \kappa_0, \sqrt{\theta_0}, \xi_0, \rho_0)$ and the parameter intervals are chosen to reflect typical values for the DAX (see Figure 8.3). In analogy to Subsection 5.2.3 we further intersect the parameter hypercube listed in Table 8.1 with the set of points satisfying the Feller condition. Note that the intervals do not encompass all observations of the parameters since this would be equivalent to an unrealistic foresight in our empirical application.

The robust hedges are constructed for an up-and-out call (UOC) with expiry $T = 1$ year, strike at 90% and barrier at 120% of the level of the underlying asset price, and a down-and-out put (DOP) with the same expiry, strike at 110% and barrier at 80%. Furthermore, we choose the initial spot $S_0 = 100\%$ and a constant riskfree interest rate of $r = 3\%$. To hedge the UOC, we make use of the ten standard calls listed in Table 8.2 which were liquidly traded on the EUREX in January 2006. In analogy we set up a portfolio of ten puts with appropriate strikes/expiries to hedge the DOP.

Table 8.2 reports the optimal portfolios and the cost corresponding to the stepwise robustifications A - C. It can be seen that (as in previous examples) the hedge portfolios essentially consist of three components. Naturally, a position in the vanilla option with the same strike and expiry as the barrier option is taken. This hedges perfectly in the absence of a knock-out event. Secondly, a large negative payoff with the same expiry as the barrier option is constructed just beyond the barrier. The negative time value of this payoff balances the positive payoff from the underlying vanilla along the barrier. Thirdly, small positions are taken in the instruments with expiry before the barrier option. These trim the balancing of the time values of the payoffs expiring at the same time as the barrier option.

Table 8.2 reveals that the composition of the hedge changes quite a bit across the steps. In particular it can be seen that the positions forming the call/put spread are reduced if we increase the robustness against changes of the volatility surface. The more protection we want to incorporate in the hedge, the less aggressive we have to choose our positions. In addition the portfolio weights of the options with expiry before the maturity of the barrier option are reduced as the robustness of the hedge increases. Among the three proposed robustification types A, B, and C, the protection against changes of the correlation ρ in step C, i.e. against changes in the skew, seems to have the most significant impact for the UOC, whereas the DOP is more affected by the volatility of variance ξ.

This differing behavior can best be explained by analyzing the impact of the Heston model parameters on the dominating digital positions at the barrier in the form of call and put spreads. For the UOC the hedge portfolio contains a large negative position in the call C^2 with strike $K_2 = 120.0\%$ and a large positive position in the call C^3 with *higher* strike $K_3 = 121.8\%$. The value of this call spread is reduced if the cost of call C^2 increases relative to the cost of call C^3, or equivalently if the slope of the implied volatility surface becomes more negative in strike direction. Hence the critical volatility surface scenarios for the static hedge of an up-and-out call are those with steep skew, which in turn corresponds to very negative values of the correlation parameter ρ. Thus a robustification against the parameter ρ can be expected to have the most significant effect on the portfolio positions as well as the hedge portfolio cost. This intuition is clearly confirmed by the results in Table 8.2. For the DOP the situation is reversed. In this case the large negative position is taken in the put with strike 80.0%, which is set off by a huge positive position in the put with *lower* strike 78.8%. Thus the critical volatility surface scenarios for the DOP are those with an almost flat[1] downside of the volatility surface in strike direction. Within the HSV model, these scenarios are best reflected by a small volatility of variance parameter ξ. Correspondingly, robustifying against changes of ξ is more important in this situation than allowing for variations of ρ. Table 8.2 illustrates these results in the form of changes of the cost and portfolio positions throughout the robustification steps A, B and C.

In summary the robust static hedge automatically guarantees the hedge performance for the critical states of the forward skew. This conservative approach can also be confirmed by comparing the cost of the static super-replication portfolios listed in Table 8.2 with the cost of dynamic hedge portfolios. For the HSV model with parameters p_0 the theoretical fair value of the UOC and DOP can be computed as $\mathrm{UOC}_{\mathrm{HSV}}(0) = 4.22\%$ and $\mathrm{DOP}_{\mathrm{HSV}}(0) = 2.55\%$. In contrast, a local volatility

[1]Equity markets are usually not faced with a positive skew. Within the HSV model, this would lead to positive correlations ρ which we explicitly excluded from the parameter hypercube described in Table 8.1.

UOC	T_i	1.0	1.0	1.0	1.0	1.0	0.75	0.75	0.75	0.5	0.5	
	K_i	90.0%	120.0%	121.8%	125.4%	130.8%	120.0%	123.6%	130.8%	120.0%	127.2%	Cost
	A	1.0	−36.09	41.70	−7.72	1.30	0.31	0.24	−0.76	0.06	0.13	4.33%
	B	1.0	−34.22	37.51	−3.72	−0.61	0.13	0.28	−0.38	0.01	0.14	4.36%
	C	1.0	−28.78	28.40	1.20	−2.09	0.19	−0.07	0.21	0.08	−0.03	4.72%
DOP	T_i	1.0	1.0	1.0	1.0	1.0	0.75	0.75	0.75	0.5	0.5	
	K_i	110.0%	80.0%	78.8%	76.4%	72.8%	80.0%	77.6%	72.8%	80.0%	75.2%	Cost
	A	1.0	−79.48	100.0	−25.87	5.87	−0.86	1.09	−0.55	0.04	−0.11	2.57%
	B	1.0	−51.59	43.36	17.06	−9.60	0.02	0.01	0.14	−0.01	0.02	3.00%
	C	1.0	−46.41	33.04	24.74	−12.37	0.04	−0.02	0.23	−0.01	0.02	3.09%

Table 8.2: Hedge portfolio specifications in the form of portfolio weights α_i for the different robustification steps listed in Table 8.1. Hedge options are calls for UOC and puts for DOP. Strikes and costs (in the sense of HSV prices) are in percent of the level of the underlying.

model calibrated to the volatility surface implied by the HSV model with parameters p_0 leads to the prices $\mathrm{UOC}_{\mathrm{LV}}(0) = 3.36\%$ and $\mathrm{DOP}_{\mathrm{LV}}(0) = 2.79\%$, respectively. The price differences illustrate the significant model risk for an up-and-out call and down-and-out put. Although both the HSV and the local volatility model are calibrated to the same volatility surface at time $t = 0$, the vastly differing model dynamics lead to very different assumptions regarding the forward skew. It is well known (see e.g. Gatheral [43]) that the forward skew is flattening quickly in local volatility models, whereas it is approximately time-homogeneous in stochastic volatility models. But this means that the digital risk of the DOP is priced more conservatively in a local volatility model ($\mathrm{DOP}_{\mathrm{LV}}(0) = 2.79\% > 2.55\% = \mathrm{DOP}_{\mathrm{HSV}}(0)$), whereas the steeper forward skew of the HSV model produces higher prices for the up-and-out call ($\mathrm{UOC}_{\mathrm{LV}}(0) = 3.36\% < 4.22\% = \mathrm{UOC}_{\mathrm{HSV}}(0)$). In comparison to these dynamic hedging approaches the robust static hedge does not make a particular assumption regarding the forward skew. Instead it robustifies the portfolio against all sorts of future volatility surfaces reflected by some parameter in the hypercube. This more conservative viewpoint comes at a fairly low cost. Table 8.2 reveals that the UOC static hedge is only 50 basis points more expensive than the HSV dynamic hedge. Similarly, the DOP robust static hedge only requires an additional 30 basis points on top of the cost of the more conservative local volatility dynamic hedge.

The arguments above illustrate that the robust static hedge is up to a certain degree model-independent. In particular the model generating the volatility surface scenarios (in our case the HSV model) does not matter. If two models parameterize the same set of volatility surfaces, the resulting static hedge will be the same. However, the Heston model is particularly convenient for the construction of the hedge since it only has a few parameters steering the shape of the implied volatility surface. The results presented in this subsection showed that in general the inclusion of all five model parameters in the robustification is necessary. Therefore, the empirical analysis is based on super-replicating hedges corresponding to robustfication step C in Table 8.2.

8.2.2 Sensitivity Analysis With Respect to Initial Cost

The robustification approach has the advantage of preserving the super-replicating property of a hedge under changing market conditions. However, the algorithm finds the cost-optimal portfolio based on the market prices observed when the hedge is constructed. These market prices can be expressed by the parameters $(\sqrt{Y_0}, p_0) = (\sqrt{Y_0}, \kappa_0, \sqrt{\theta_0}, \xi_0, \rho_0)$ of the HSV model calibrated at time $t = 0$. But this means that different market prices (or parameters p_0) in general lead to different super-replication hedges for barrier options with identical specifications set up on different days even when the hedges are robustified against the same parameter hypercube. Thus, some of the appeal of the robust approach is lost since dependence on the current calibration still lingers. This observation raises questions about the sensitivity of the robust static hedge against changes of the initial parameter vector $v_0 = (\sqrt{Y_0}, p_0)$.

In the following we analyze this question for the hedge of the UOC listed in step C in Table 8.2. To be more precise, we fix the parameter hypercube corresponding to step C in Table 8.1 and recompute the optimal hedge for the UOC using random initial parameter vectors, $v_i, i = 1, \dots, 200$, chosen uniformly in the same parameter hypercube. It is clear that all these hedge portfolios have the super-replication property as long as the market can be represented by a vector from the hypercube. It is also clear that if the current market prices were reflected by the initial parameter v_i, then the optimal hedge corresponding to v_0 must be at least as expensive as the optimal hedge computed for the initial parameter vector v_i. What remains to judge is the magnitude of the sub-optimality cost of the v_0-hedge in this situation.

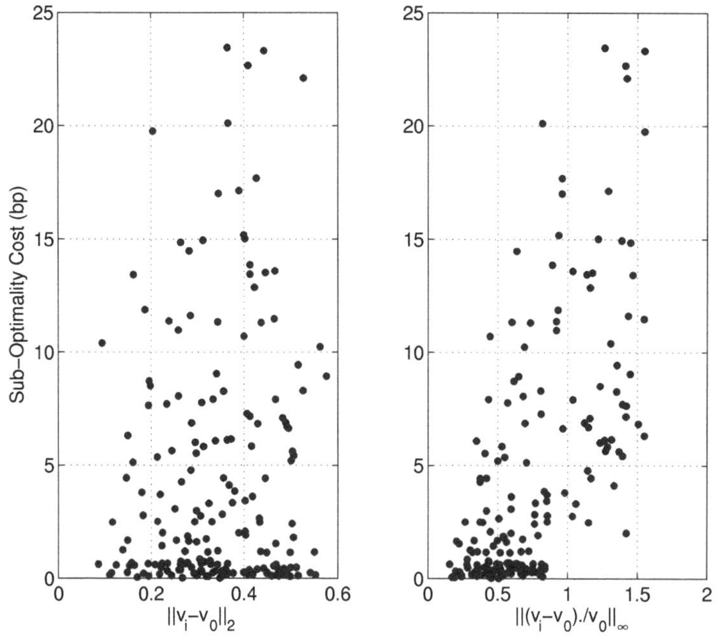

Figure 8.4: Cost in basis points of using a minimal assumptions hedge of the UOC, based on v_0, when market conditions, described by v_i, are drawn uniformly from the parameter hypercube in Table 8.1. The riskfree rate was set to $r(t) = 3\%$.

In Figure 8.4 the cost of the sub-optimality in basis points, $\text{Cost}_0 - \text{Cost}_i$, based on market prices reflected by v_i, is shown. The left panel shows the cost as a function of the overall distance between the fixed v_0 and the random v_i. The right panel shows the cost as a function of the largest relative parameter component deviation from the component of v_0. From the graphs it can be seen that although the parameter hypercube represents quite diverse market conditions the cost of sub-optimality is (often much) smaller than 25 basis points. The mean cost of sub-optimality is a mere 4 basis points. It is not clear that a larger overall difference between two market scenarios, $||v_i - v_0||_2$, leads to larger cost. What determines the magnitude of the cost rather seems to be the relative deviation in the value of a parameter.

The previous observations allow to conclude that the sub-optimality of using a fixed parameter vector in changing market conditions only carries a small cost. The average sub-optimality of 4 basis points shows that the initial market prices only have a small effect on the robust static hedge if compared to the effect of varying the parameter hypercube. This allows us to neglect changes of the market prices and instead choose a fixed parameter $v_0 = (\sqrt{Y_0}, p_0)$ for the computation of the hedge throughout the considered dataset. The implications of this are far-reaching. Fixing the parameter v_0 in problem (5.9) implies that we have a hedge portfolio where the portfolio weights are fixed even though the market changes. This is possible because the optimization problem (5.9) is homogeneous in the asset level when all contract levels are specified relative to the current asset level. The resulting hedge portfolio retains the super-replicating property as long as the market can be described by a vector in the hypercube, that is the hypercube is the only assumption we make about the parameters. Therefore, we refer to the portfolios listed in step C in Table 8.2 as minimal assumptions (MA) hedge portfolios. These MA hedges will allow us in Section 8.5 to set up static hedge portfolios with *the same* portfolio weights for barrier options issued on each day of seven years of data.

8.3 Other Hedging Approaches Used in the Study

An empirical evaluation of the performance of the robust super-replicating hedge is interesting in itself. However, to judge the performance of the robust hedge we chose to compare it with two replicating hedge strategies – a standard dynamic hedge as well as a replicating static hedge. For the dynamic hedge a large variety of models are available. In this study we choose a dynamic hedge based on the local-volatility model, because this kind of models is frequently used in practice and differs from the HSV model we used to set up the static hedge.

The hedge performance of the local volatility (LV) model for the reverse barrier problem is thoroughly studied in Engelmann, Fengler and Schwendner [33]. In that paper a number of variants of the LV model as it is used by practitioners are studied. The focus is on the socalled 'stickiness'. When the sensitivity of the option price with respect to movements in the underlying is computed three methods are commonly used in practice. First, one can assume that the local volatility surface remains fixed. This is consistent with the LV model, but not with the observed market behavior. Second, one can assume that the implied volatility surface remains fixed. This is called the sticky-strike approach. Finally, one can assume that the implied volatility surface floats with the price of the underlying. This is called the sticky-moneyness approach. The stickiness assumptions are ad-hoc ways to capture the dynamics of the implied volatility surface. One main result of [33] is that the sticky-strike approach is slightly better than the other approaches. Furthermore, a hedge based on both the option delta and vega is found to be considerably better than a pure delta hedge and only slightly less effective than a delta-vega-vanna hedge.

The dynamic hedge we consider is a sticky-strike delta-vega hedge (DV) in the LV model, that is we hedge both against shifts in the underlying and (parallel) shifts of the implied volatility surface. To compute the vega the implied volatility surface is shifted and the sensitivity is found by recalibration and a finite difference approximation. The option sensitivities are hedged by taking positions in the money market account, the spot and the ATM vanilla option with the same expiry as the barrier option. The hedge strategy is rebalanced daily.

The second alternative hedge strategy is the strike-spread approach (STR) of Carr and Chou [23] as well as Carr, Ellis and Gupta [24] presented in Subsection 2.3.1. Although the STR hedge is static like the MA hedge introduced in Subsection 8.2.2, it is neither super-replicating nor robust with respect to the model parameters. From this point of view it will be interesting to compare its empirical performance to the robust static superhedge.

8.4 Experimental Design

To evaluate the performance of a hedge strategy one should focus on the hedge errors and particularly their dispersion. Thus, time series of realized hedge errors are needed. This section describes the way in which time series of realized hedge errors are constructed.

For each day we initiate a reverse up-and-out call and a reverse down-and-out put. That is, we tackle hard hedging problems posed by single barrier options. For each contract we set up the robust MA hedge, a static STR hedge as well as the dynamic DV hedge. The hedges are monitored daily with the dynamic hedge being rebalanced using a recalibrated LV model. If the rebalancing of the dynamic hedge is not self-financing the money market account is adjusted to reflect this. The money market account is rolled at the short-term interest rate of the relevant day. A hedge error, that is the difference between the payoff from the hedge and that of the corresponding barrier option at knock-out or expiry is recorded. This is done for each of the first 1500 days in the sample which means that all barrier options may run to expiry in the absence of a knock-out event.

This design results in time series of realized hedge errors using market data for each of the three hedges and each barrier option type (UOC/DOP). To make the hedge errors comparable, all levels referring to the DAX are in percent of the DAX at initiation of the particular barrier option. As in Section 8.2, the UOC has strike at 90% and a barrier at 120% of the DAX. For the DOP we chose a strike and barrier of 110% and 80%, respectively. Correspondingly, the STR hedge uses as hedge options calls with strikes at $\{90\%, 120\%, 121.8\%, 160\%\}$ in the UOC case (see Subsection 2.3.1), whereas puts with strikes $\{110\%, 80\%, 78.8\%, 58.2\%\}$ are used for the DOP hedge. The MA hedge is given by the portfolio weights and options with the strikes and maturities listed in Table 8.2 for robustification step C.

When closing a hedge, the hedge error is rolled at short-term interest rates until expiry. If no knock-out event occurs for a given barrier option, then the hedge is unwound at observed market prices and the hedge error is recorded. In case the barrier is hit, a choice must be made regarding the closing of the hedge. One possibility is to unwind the hedge at the observed spot beyond the barrier. This is consistent with the data, but inconsistent with the hedge strategies since these require the hedge to be closed exactly when the knock-out event occurs. Therefore, we also consider closing the hedges exactly at the barrier but assuming that the observed implied volatilities apply at this time. Effectively, the latter approach shifts the spot from the observed level to the barrier. This is (often slightly) inconsistent with the data, but consistent with the hedge strategies. A point in favor of choosing the first approach is that it is unlikely that a hedge can be unwound exactly at the knock-out event in practice. Using the spot beyond the barrier can be seen as using a proxy for this liquidity risk. Closing the hedges on the barrier however focuses on the skew sensitivity of the barrier options. Hence, both approaches seem relevant and we report results for both.

8.5 Empirical Results

To be able to properly interpret the time series of hedge errors, information on the knock-out events must be considered. The number of knock-out events as well as the average life time of both types of barrier options are reported in Table 8.3. From the table it can be seen that the DOP case will be the hardest to hedge. More DOP than UOC contracts knock-out after a slightly shorter average life time. Furthermore, the moves beyond the barrier are larger on average in the DOP case.

Option	Knock-out	Avg. Life Time	Avg. Over-/Undershoot
UOC	39.5%	140	0.50%
DOP	48.2%	116	−1.78%

Table 8.3: Proportion of 1500 barrier options that knocked out. The average life time is in days and the average over- and undershoots are relative to the barrier.

The first question to consider is whether the super-replicating property of the robust hedges holds empirically. To answer this, the realized hedge errors from both hedge closing approaches for the UOC case are shown in Figure 8.5. It can be seen that the super-replicating property is essentially confirmed empirically. This holds for both hedge closing approaches. There are 3 observations of negative hedge errors, which are all smaller than 30 basis points in absolute value. This implies that the market conditions prevailing at the knock-out dates correspond to some parameter vector in the hypercube, at least for the hedge options.

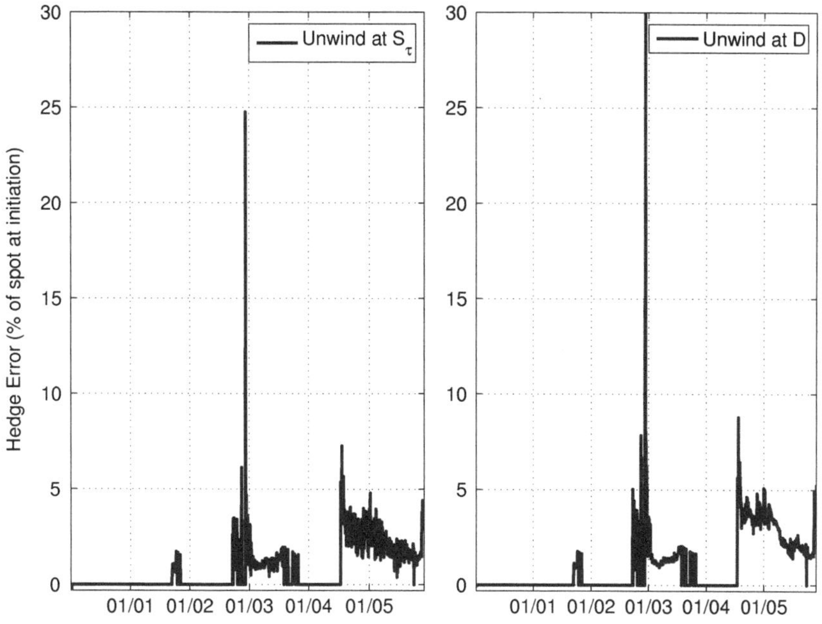

Figure 8.5: Time series of hedge errors for the UOC case. Series for both unwinding at the observed spot S_τ at the time of the barrier hit, and at the barrier D are shown.

Similarly, Figure 8.6 shows the hedge error time series for the DOP case. In this case, the super-replicating property is found to hold only in the case where the hedges are closed on the barrier. For the data consistent hedge closing approach there are 38 negative hedge errors 6 of which are below -4% which is approximately the size of the mean hedge cost. The three very large hedge errors occur when the knock-out event happens close to expiry and the spot moves to about 78% of spot at initiation, that is just beyond the barrier. In these cases, the large short position in the 80% put option dominates the payoff. Relative to the sample size of 1500 contracts only a small number generates negative hedge errors. A few of these however are spectacular. Thus, the conclusion in the DOP case is that the robust static portfolio succeeds in hedging the skew risk, but that the portfolio is still exposed to liquidity risk. This is not a surprise regarding the increased liquidity risk for the DOP case illustrated in Table 8.3 as well as the fact that optimization problem (5.9) by definition assumes the liquidation of the hedge portfolio on the barrier.

Although it is in principle possible to additionally robustify the robust static hedge against such liquidity/gap risk (see Section 6.2), we do not pursue this approach here, because we did not apply any similar techniques for the DV and STR hedge. In reality,

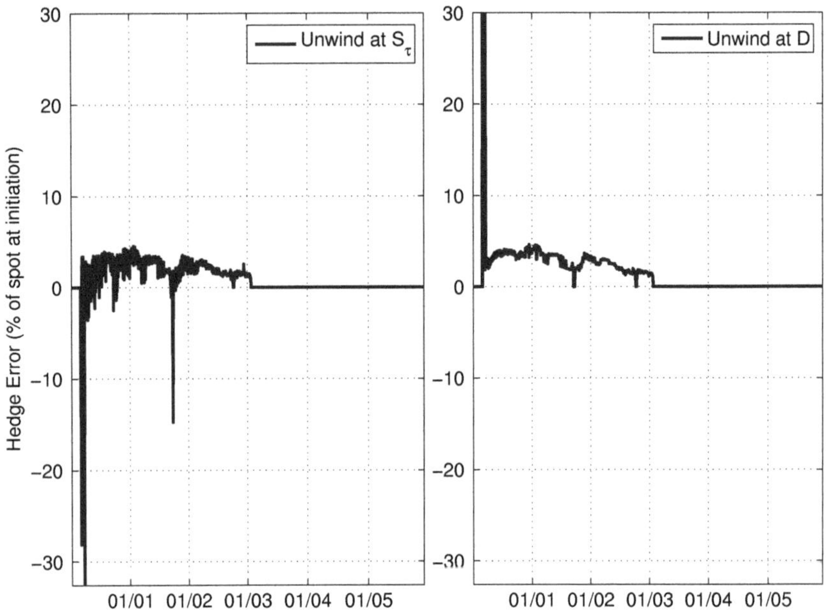

Figure 8.6: Time series of hedge errors for the DOP case. Series for both unwinding at the observed spot, S_τ, and at the barrier, D, are shown.

it is reasonable to expect the MA hedges to be closed somewhere between the two hedge closing approaches we consider. This will tend to make the negative hedge errors less frequent. Finally, note that static hedges with an STR-like component, such as the STR and MA hedges are relatively insensitive to large negative moves, such as the 8.89% drop on 11 September 2001. If the hedge is closed at the closing price, then the intrinsic value of the hedge is close to zero, since the spot has leaped the strike-interval where the hedge has a large negative payoff.

In the rest of this section we focus on the errors generated by unwinding at the observed spot, that is, using the data consistent approach. The reason for this is that we believe liquidity risk could well be an issue when attempting to unwind potentially large vanilla positions. Furthermore, this approach does not favor the robust MA hedge unjustly relative to the other hedging approaches we consider.

Next, the frequencies of hedge errors corresponding to the MA, STR and dynamic hedges of the UOC are compared in Figure 8.7 (left). From the graphs it can be seen that a main advantage of static hedges with an STR-like component is that a perfect hedge is obtained in the absence of a knock-out event. This is not the case for dynamic hedges where the hedge portfolio must be closed at market prices at expiry. This difference explains the large peak at zero hedge error shared by the MA and STR hedges. The near-perfect super-replication of the robust MA hedge is found as before. For the STR hedge the interval where the hedge has a large negative intrinsic value is considerably wider than for the MA hedge. This, as well as the lack of positive payoff beyond the barrier and the presence of an implied volatility skew, leads to negative STR hedge errors in almost all knock-out events. The hedge errors for the MA and STR hedges are almost mirror images of each other reflected through zero. The hedge error distribution of the DV hedge on the other hand is uni-modal with a fat positive tail. It is noteworthy that the dynamic hedge generates a larger number of positive hedge errors than the MA hedge, even though the MA hedge was constructed without direct penalty for large infrequent hedge errors.

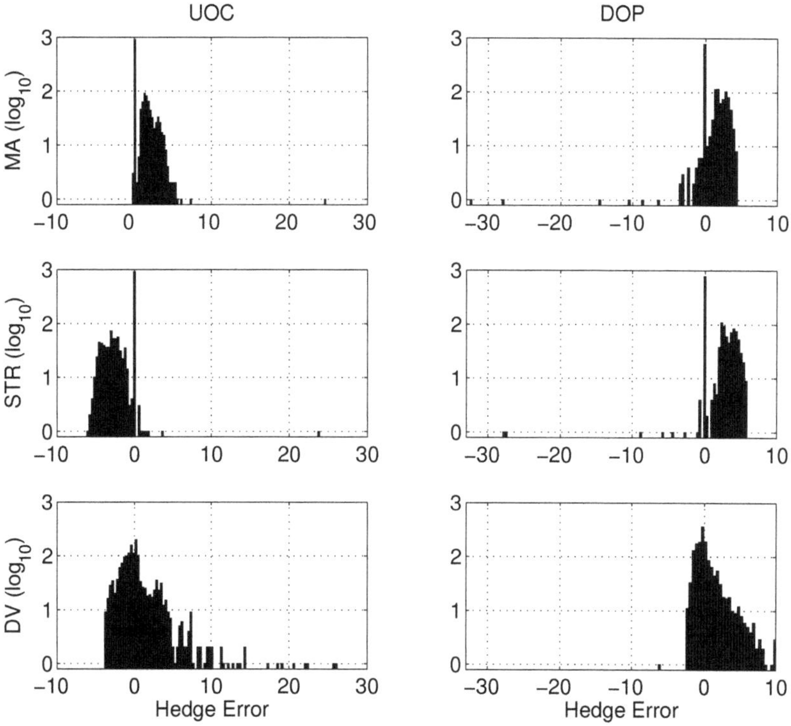

Figure 8.7: Hedge error frequencies for the MA, STR and DV hedges of UOC and DOP contracts. Hedge errors are in percent of spot at initiation.

Similarly, the frequencies of hedge errors for the three hedge strategies in the DOP case are shown. Again, the peak at zero hedge error is shared by the MA and STR hedges corresponding to a perfect hedge when there is no knock-out event. The MA hedge still has mainly positive hedge errors, but the negative errors discussed above are visible. Apart from the shared peak at zero, the positive hedge errors from the MA hedge are closer to zero than those from the STR hedge. This shows the strength of the approximate cost-optimality used when constructing the MA hedge. The majority of the nonzero hedge errors from the STR hedge are positive, which can be explained with the presence of an implied volatility skew compared to the model assumptions on whose basis the hedge was constructed. In contrast to the MA and STR hedge, the DV hedge avoids the spectacular negative errors.

Summary statistics for the different hedges are presented in Table 8.4. The sensitivity of the results for the STR hedge emphasizes the fragility of this particular hedging strategy. The lack of consistent performance of STR-like hedges was one of the observations in Engelmann et al. [32]. The magnitude of the results gives an indication of the performance of the MA hedge. Apart from the preservation of the super-replicating property discussed above, the focus is on the dispersion and cost of the MA hedge. The measure of dispersion we use is mean absolute deviation about the median, 'mad.'. The result is that the MA hedge has the lowest dispersion for both barrier option types, particularly in the UOC case. This result is surprising since we actually used *the same* portfolio weights for each MA hedge set up within the 1500 day sample.

	Hedge	cost	P(Loss)	min	max	mean	med.	mad.	skew.	kurt.
UOC	MA	5.30	0.20	−0.3	24.8	0.8	0.0	0.8	5.0	75.9
	STR	3.48	38.80	−6.2	24.0	−1.1	0.0	1.2	1.1	29.5
	DV	3.94	55.67	−3.9	26.2	0.3	−0.2	1.6	3.6	24.7
DOP	MA	3.38	2.53	−32.6	4.5	0.9	0.0	1.1	−7.0	119.0
	STR	3.88	0.80	−28.0	5.9	1.6	0.0	1.7	−3.0	45.1
	DV	2.69	60.53	−6.4	35.8	0.3	−0.2	1.2	5.5	59.1

Table 8.4: Summary statistics for all hedges in both the UOC and DOP cases. The mean absolute deviation about the median is labelled 'mad.'. All numbers except skewness, kurtosis and the loss probability are in percent of spot at initiation. The cost is computed from market prices.

In the UOC case the cost, computed from market prices, of the MA hedge on first sight seems to be significantly higher than the other hedges. However, recalling our analysis from Subsection 8.2.1, the increased cost is caused by differing assumptions regarding the dynamics of the volatility surface. The quickly flattening forward skew of the LV model prices the digital risk contained in the UOC less conservatively than the MA hedge. Table 8.4 reveals that the more conservative model assumptions of the MA hedge also lead to much more conservative UOC hedge error statistics compared to the other hedging strategies. However, the ultimate decision for one hedge or the other basically depends on the degree of risk aversion of the hedger. In this context we mention that the cost of the MA hedge would in fact be closer to the cost of a (more conservative) dynamic hedge in a stochastic volatility model. In contrast to this, in the

DOP case the DV hedging cost and the price of the MA hedge are much closer to each other, because in this situation a flat forward skew is the more conservative scenario.

A final angle to consider is the bounds sub- and super-replicating hedges put on fair barrier option prices. Since we have near-perfect super-replication the cost of the MA hedge provides an upper bound on the fair prices of UOC and DOP contracts, respectively. Similar to the super-replicating version, one can also construct a sub-replicating minimal assumptions hedge (see Section 6.1) using the parameter values given in Table 8.1. For the sake of brevity we summarize the results in Table 8.5. The sub-replicating hedges achieve true sub-replication in the sample as can be seen from the maximum hedge errors. The cost bounds set by the MA hedges are [3.50%, 5.30%] for the UOC and [1.38%, 3.38%] for the DOP. These bounds are wider than those listed in Table 6.3 since the price intervals are not tailored to an individual barrier option, but instead correspond to 1500 contracts. Note that the cost 3.94% of the DV hedge for the UOC is not too far away from the lower bound of the corresponding interval, which indicates that this particular strategy is an aggressive one for up-and-out calls. On the other hand, comparing the DOP cost of 2.69% of the DV hedge with the robust interval confirms that the hedge is much more conservative for down-and-out puts.

	cost	$P(\text{Gain})$	min	max	mean	med.	mad.	skew.	kurt.
UOC	3.50	0.00	−12.5	0.0	−1.2	0.0	1.2	−1.8	6.0
DOP	1.38	0.00	−157.0	0.0	−3.0	0.0	3.0	−10.1	152.8

Table 8.5: Summary statistics for sub-replicating MA hedges in both the UOC and DOP cases. The mean absolute deviation about the median is labelled 'mad.'. All numbers except skewness, kurtosis and the gain probability are in percent of spot at initiation. The cost is computed from market prices.

In summary the results show that the robust static hedge has a quite impressive performance at an acceptable additional cost. The empirical analysis confirms that in contrast to typical dynamic hedging strategies the robust static super-replicating hedge neither takes a view on the forward skew nor on the volatility surface dynamics. Instead the minimal assumptions hedge only depends on the barrier option specifications and market-typical bounds on model parameters. As a consequence the robust static hedge is mainly depending on the particular underlying of the barrier option. The more wildly the volatility surface of the underlying moves through time, the more expensive the hedge is. However, compared to typical replicating hedges and regarding the degree of robustness the cost is found to be quite low. Finally the minimal assumption hedges have the favorable property that the portfolio weights stay the same, even if the hedges correspond to barrier options issued on different days and in different market environments.

The empirical analysis revealed that the super-replicating property of the minimal assumptions hedge was essentially found to hold for the data sample in the UOC case.

The DOP case showed more negative hedge errors, a few of which were spectacular, but not greater than those incurred by other static hedging strategies. The results reflect the fact that hedging of reverse barrier options is a hard problem, but the dispersion of hedge errors for the robust hedge is smaller than those of the considered alternative hedging approaches in both the UOC and DOP cases. Near-perfect super-replication and small dispersion comes at a fairly low extra cost. This result is not only in favor of the robust hedge but also of the minimal assumptions version. Finally, constructing sub-replicating minimal assumptions hedges leads to empirically robust price bounds for the barrier options on the DAX that were considered.

The presented results motivate further empirical research on the performance of gap-robustified static and dynamic hedges. The theoretical foundation for a gap-robust version of the robust hedge is found in Section 6.2. The ad-hoc practice of 'moving the barrier' when hedging barrier options is the corresponding approach for dynamic hedges. Clearly, hedge portfolios for down-and-out puts would benefit from such a robustification. Moreover, it would be interesting to compare the hedge performance of the minimal assumption hedge and the delta-vega hedge in a local volatility model to a suitable dynamic hedge derived in a stochastic volatility model. Regarding the differing forward skew dynamics such a hedge can be expected to be closer to the MA hedge in the UOC case, whereas it should produce less conservative portfolios for the DOP.

9 Summary and Outlook

In this work we developed a new static hedging approach for barrier options which is based on the idea of super-replication. We showed that this concept leads to a stochastic optimization problem whose numerical solution results in the cheapest hedge portfolio consisting of a given set of liquidly traded standard options. As it turned out, the computed hedging strategies are only marginally more expensive than the barrier option itself and fit the delta and vega of the target option closely.

Moreover, the special structure of the stochastic super-replication problem allowed to transform it into a deterministic linear semi-infinite optimization problem which can be solved by efficient numerical methods and reduces the computation time in comparison to Monte Carlo-based algorithms from several minutes or even hours to a few seconds. In addition the deterministic problem representation was used to theoretically derive the dual optimization problem which maximizes the expected discounted barrier option payoff over the set of price-consistent probability measures. These findings demonstrate that static hedge portfolios for barrier options are closely related to measures providing a static fit of the volatility surface. Surprisingly, it was also possible to prove the existence of an optimal price-consistent measure which is the finite weighted sum of Dirac measures whose support corresponds to the active super-replication constraints.

But the relevance of the volatility surface for static hedge portfolios for barrier options is not only reflected by the structure of the dual problem. By definition the portfolios consist of calls and/or puts of several strikes and maturities. Hence static hedge portfolios are extremely sensitive to the smile assumption of the model used for the derivation of the strategy. Numerical results confirmed that volatility shocks or changes of the skew can lead to extreme hedging losses amounting to multiples of the fair value of the barrier option. However, while this also applies to all other static hedging approaches in the literature (neglecting the too conservative model-independent hedge of Brown, Hobson and Rogers [19]), the flexibility of the static super-replication framework allowed for the first time to rigorously incorporate robustness against changes of the volatility surface into the design of the static hedging strategy. This was achieved by accounting for the model parameter uncertainty within the semi-infinite optimization problem in the sense of a worst case design. Based on Mizohata's uniqueness theorem for parabolic partial differential equations we were able to prove the existence of solutions of the robust optimization problem as well as the convergence of suitable semi-infinite optimization algorithms to a robust trading strategy in the continuous time financial market model. Numerical results for the Black–Scholes and Heston's stochastic volatility model demonstrated that the risk of a changing volatility surface can be eliminated by surprisingly low cost without increasing the number of options in our portfolio.

Furthermore, the theoretical and numerical results carried over in analogy to the static sub-replication problem which yields a lower bound for the price of the target option. Combined with the super-replication price, this results in a price interval for the barrier option that is surprisingly tight regarding the achieved robustness of the trading strategies.

A further analysis showed that jump risk or liquidation delays on the barrier are an additional source of uncertainty which may cause huge hedge errors if the barrier is hit close to maturity of the barrier option. But again, this uncertainty was eliminated by an appropriate modification of the static super-replication problem. Moreover, we were able to reduce the resulting robust optimization problem to a much simpler problem which can be interpreted as moving the barrier.

Afterwards we investigated the question of potential model errors. Once more we were able to robustify the static hedging strategies against this form of uncertainty by allowing the real world prices to deviate from the model prices within prespecified ellipsoids in the size of a few basis points. Although this significantly increased the complexity of the static hedging problem, it turned out that after some suitable transformations a solution of the problem can be computed by solving a sequence of nonlinear and second order cone problems or semidefinite and nonlinear semidefinite programming problems.

It should be mentioned that these findings are not limited to a particular barrier option, but instead transfer in analogy to up/down-and-in/out options with or without rebates, with constant, time-dependent or discrete barrier or even double knock-out feature. The desired degree of robustness for any of these products can be achieved by eliminating the individual risk components (volatility shocks, skew risk, stock price jumps, model errors) from the trading strategy.

An empirical analysis based on a seven year dataset covering various interesting market scenarios confirmed the robustness of the new static hedge against changes of the volatility surface (including changes of the skew). Compared to the strike-spread static hedge and a dynamic local volatility hedge, the robust superhedge additionally lead to the lowest hedge error dispersion. This result is in particular surprising since we set up the same robust static hedge (in the sense of portfolio weights) for all barrier options issued throughout the dataset.

Finally, to illustrate the conceptual difference of the existing static hedging approaches in the literature and the proposed static super-replication framework, we briefly summarize the main ideas. The existing approaches can be classified into analytical results and discrete replication approaches. On the one hand, the analytical results of Carr et al. (see [23] and [24]) aim at perfectly replicating the value of a barrier option in a given financial market model. To achieve this, it is essential to assume the availability of hedge instruments with specific strikes whose tradability is questionable, as well as simple model dynamics which allow theoretically exact derivations. But these derivations fail under the assumption of realistic smile dynamics. Consequently, the analytically derived static hedging strategies can lead to large hedging losses if the imposed assumptions are dropped. On the other hand, the discrete replication approaches of Derman, Ergener and Kani [27], Fink [38], Nalholm and Poulsen [73] as well as Allen and Padovani [2] construct a static hedging strategy by first fitting the payoff of the barrier option at maturity, and afterwards exactly matching the payoff along the barrier for a given finite number of scenarios. These scenarios might simply be different points in time at which the barrier may be hit (Derman et al.), time-volatility combinations (Fink), time-volatility-stock price combinations (Nalholm and Poulsen) or even different time-volatility-skew scenarios (Allen and Padovani) at the time of the barrier hit.

The results in this book showed that a sufficiently robust static hedge portfolio has to consider volatility shocks, skew risk, stock price jumps and barrier hits at any time - otherwise huge hedge errors might occur. To exactly fit the barrier option in all four problem dimensions requires a scenario grid in a multidimensional space whose number of nodes explodes. But the mentioned discrete replication approaches require one additional hedge instrument for each node in the mesh which can easily amount to hundreds of options even if only a few scenarios are considered. Moreover, even if hundreds or thousands of options are included in the portfolio, the residual risk for huge hedging losses is still very high, because the spacing between the scenario grid points is too large. Furthermore, the concept of exactly fitting the barrier option on the resulting grid is problematic, because the linear system to set up the portfolio approaches singularity and the task of exactly fitting a barrier option with discontinuous payoff by a portfolio of continuous call price functions is numerically ill-posed. In summary the discrete replication approaches developed in the literature suffer of the following problems:

1. Ill-posedness
2. Only a few scenarios can be considered which implies high residual risk
3. The portfolio consists of a huge number of standard options (up to several hundred, depending on the number of scenarios)

In contrast to this the robust super-replication approach presented in this book

1. Is numerically stable
2. Takes an infinite number of scenarios into account (preventing residual risk)
3. Is only based on a few standard options (independent of the number of scenarios).

Furthermore, the method is proven to converge to an existing robust portfolio in the continuous time financial market model. Note that these advantages are intrinsic properties of the proposed semi-infinite optimization method and not limited to the general stochastic volatility model we used throughout the book. The model was only abused to generate volatility surface scenarios – hence we might replace it with any other parametrized model.

Of course the performance of the trading strategies crucially depends on the actually chosen degree of robustness which is reflected by the model parameter uncertainty sets. The necessary size of these sets depends on the stability of the implied model parameters over time and hence the shape of the volatility surface. However, this shape may be much more stable for the relevant part of the surface containing the calls which are actually part of the hedge portfolio than for the whole set of market prices. Thus it seems desirable to analyze the stability of the implied model parameters if the model is solely calibrated to the interesting part of the volatility surface.

In addition it should be investigated if the maximum principle for parabolic partial differential equations can also be used to eliminate the volatility-dimension from the semi-infinite super-replication problem for stochastic volatility models. Similar to the robustification against stock price jumps, the maximum principle may imply that the minimal and maximal hedge error always occur on the boundary of the volatility uncertainty set. This would also support the observation in various simulation-based studies, that the maximal hedge error always occurs for barrier hits in low volatility states shortly before maturity of the barrier option.

Furthermore, it seems worthwhile to identify static hedge portfolios for barrier options by minimizing other, less conservative one-sided hedge error measures. For example one could search the cheapest portfolio subject to an expected shortfall constraint or by asking for a super-replication with a prespecified probability. However, for these extensions we will loose the equivalence of the stochastic optimization problem and a suitable deterministic problem, which was one of the key advantages of the static super-replication hedge. Consequently, a solution of such problems will most likely fall back to a Monte Carlo-based procedure in the spirit of Section 3.2, leading to a drastic increase of computation time. While it should still be easily possible to solve the non-robust problem, the increased computational cost prevents the numerical solution of the robust counterpart that also takes changes of the volatility surface into account. In this situation it might be beneficial to derive a hedge for a general one-sided error measure based on worst-case model parameters computed with the robust static super-replication method.

Finally, another natural extension is to apply the static super-replication approach to the hedging of other options. As mentioned in Section 6.3, the developed semi-infinite programming approach is a conceptual method that is applicable to the derivation of robust hedging strategies for any option which can be weakly replicated in the sense of Joshi [55]. Moreover, it seems feasible to recursively compute superhedges for feeble hedgeable exotics like Asian options.

A General Existence Theorem

The question of existence of a cost-optimal super-replicating strategy is a very general one. In the case of dynamic strategies this question has already been analyzed in depth, see e.g. [30]. However, in the case of static hedging strategies the answer is not clear yet. In this section we will prove the existence of optimal static super-replicating strategies in general financial market models under very mild conditions. Regarding the model and the traded financial instruments, we impose the following assumptions.

Assumption A.1. Let $\mathcal{M}$ be a continuous or discrete financial market model with time index set $I = [0, T]$ or $I = \{t_0, \ldots, t_f\}$, $0 = t_0 < t_1 < \ldots < t_f = T$, corresponding probability space $(\Omega, \mathcal{F}, P)$ and filtration $(\mathcal{F}_t)_{t \in I}$, $\mathcal{F}_0$ trivial. Further let $C^1, \ldots, C^n$ be tradable financial instruments in $\mathcal{M}$ with value process $(C_t^i)_{t \in I}$ adapted to the filtration and $C_t^i \in L^1(P)$ $\forall\, t \in I$, $i = 1, \ldots, n$. Finally assume that there exists an index $j \in \{1, \ldots, n\}$ with $C_t^j > 0$ (a.s.) $\forall\, t \in I$.

Our goal is to hedge a European claim $C \geq 0$ (a.s.), $\mathcal{F}_T$-measurable, $P(C > 0) > 0$, with expiration T by a portfolio of the financial instruments $C^1, \ldots, C^n$ traded in $\mathcal{M}$. However, instead of searching for the cheapest super-replicating strategy in the set of all strategies we restrict ourselves to a set of static strategies. Hence mathematically, a cost-optimal static super-replicating strategy is the cheapest strategy in a specific finite dimensional subspace $\mathcal{S}$ of the set of all trading strategies. As an example consider the set of static hedging strategies designed to hedge an up-and-out call as defined in Section 2.2 or the set of purely static strategies given by constant portfolio positions. Both sets are isomorphic to $\mathbb{R}^m$ for some $m \in \mathbb{N}$. This motivates the following assumption.

Assumption A.2. Assume that a set $\mathcal{S}$ of trading strategies $(\phi_t^1, \ldots, \phi_t^n)_{t \in I}$, $|\phi_t^i| < \infty$ (a.s.) $\forall t \in I$ $\forall i$, where ϕ_t^i denotes the quantity of financial instrument C^i at time t, is isomorphic to $\mathbb{R}^m$ for some $m \in \mathbb{N}$.

Due to the isomorphism each strategy $(\phi_t^1, \ldots, \phi_t^n)_{t \in I} \in \mathcal{S}$ is uniquely identified by some $\alpha \in \mathbb{R}^m$. Hence the value of the portfolio at time t, given by $\Pi_t := \sum_{i=1}^n \phi_t^i C_t^i$, solely depends on α such that $\Pi_t = \Pi_t(\alpha)$. Furthermore the isomorphism implies that $\Pi_t(\cdot)$, mapping from $\mathbb{R}^m$ to some space of random variables, is linear. Equipped with this notation, we can define the concept of cost-optimal super-replicating strategies in $\mathcal{S}$ in a very general setting.

Definition A.3. Suppose that Assumptions A.1 and A.2 hold. Let the value of a portfolio of the financial instruments $C^1, \ldots, C^n$ at time $t \in I$ be given by $\Pi_t(\alpha)$, where $\alpha \in \mathbb{R}^m$ denotes the unique vector corresponding to some strategy $(\phi_t^1, \ldots, \phi_t^n)_{t \in I} \in$

$\mathcal{S}$. Further let the European target claim $C \geq 0$ (a.s.) be $\mathcal{F}_T$-measurable and satisfy $P(C > 0) > 0$. Then a "cost-optimal super-replicating $\mathcal{S}$-strategy" is defined as a solution (if existent) of the stochastic optimization problem

$$\begin{aligned} &\min_{\alpha \in \mathbb{R}^m} \Pi_0(\alpha) \\ &\text{s.t. } \Pi_T(\alpha) \geq C \text{ (a.s.).} \end{aligned} \tag{A.1}$$

The set of feasible points shall be denoted by $\mathcal{SR} := \{\alpha \in \mathbb{R}^m : \Pi_T(\alpha) \geq C \text{ (a.s.)}\}$.

A simple analysis shows that optimization problem (A.1) is the minimization of a linear function on a closed, convex set. If the set $\mathcal{SR}$ would be bounded, then the question of existence of a solution would be trivial. However, for $\alpha \in \mathcal{SR}$ and all $\lambda \in [1, \infty)$ we have that $\Pi_T(\lambda\alpha) = \lambda\Pi_T(\alpha) \geq \lambda C \geq C$ (a.s.). Hence $\mathcal{SR}$ is unbounded and the existence of a solution cannot be derived that easily. In particular $\mathcal{SR}$ does not have to be of polyhedral structure such that the usual linear programming arguments do not hold. Nevertheless Theorem A.4 states some very mild conditions under which we can derive an existence result.

Theorem A.4 (Existence Theorem). *Suppose that Assumptions A.1 and A.2 hold and that Π_t and C are given as stated in Definition A.3. If the market model $\mathcal{M}$ is arbitrage-free in the set of trading strategies $\mathcal{S}$ and if $\mathcal{SR}$ is non-empty, a cost-optimal super-replicating $\mathcal{S}$-strategy exists.*

Proof. Optimization problem (A.1) is the minimization of a proper convex function on the non-empty closed convex set $\mathcal{SR}$. Thus, according to Theorem 1.10, a sufficient condition for the objective to attain its infimum on $\mathcal{SR}$ is, that Π_0 and $\mathcal{SR}$ have no direction of recession in common. In the following we will characterize the common directions of recession and prove that either there are none of these directions or the optimization problem can be reduced in dimension until all common directions are eliminated.

As $\Pi_0(\cdot) : \mathbb{R}^m \to \mathbb{R}$ is a linear function, there exists a vector $b \in \mathbb{R}^m$ such that $\Pi_0(\alpha) = b^{\mathrm{T}}\alpha$ for all $\alpha \in \mathbb{R}^m$. Now assume that $d \in \mathbb{R}^m$ is a direction of recession of $\mathcal{SR}$ and Π_0, i.e.

$$\begin{aligned} &b^{\mathrm{T}}d \leq 0 \\ &d \neq 0,\ \alpha + \lambda d \in \mathcal{SR}\ \forall \lambda \geq 0\ \forall \alpha \in \mathcal{SR}. \end{aligned} \tag{A.2}$$

In case $b^{\mathrm{T}}d < 0$, we can, for fixed $\alpha \in \mathbb{R}^m$, choose $\lambda \in [0, \infty)$ large enough such that $\Pi_0(\alpha + \lambda d) = b^{\mathrm{T}}\alpha + \lambda b^{\mathrm{T}}d < 0$. Due to condition (A.2) we have $\alpha + \lambda d \in \mathcal{SR}$ such that $\alpha + \lambda d$ is a strategy with non-negative terminal payoff and negative initial cost. Obviously, this implies arbitrage opportunities which contradicts the assumption.

Thus the case $b^{\mathrm{T}}d < 0$ cannot occur and hence $d \in \mathbb{R}^m$ is a common direction of recession of Π_0 and $\mathcal{SR}$ if and only if

$$b^{\mathrm{T}}d = 0,\ d \neq 0$$
$$\alpha + \lambda d \in \mathcal{SR}\ \forall \lambda \geq 0\ \forall \alpha \in \mathcal{SR}. \tag{A.3}$$

In other words, any common direction of recession of Π_0 and $\mathcal{SR}$ is a direction in which Π_0 is constant. If $\mathcal{SR}$ is polyhedral, this immediately implies the existence of a solution of the minimization problem. However, in general $\mathcal{SR}$ might not be polyhedral such that we cannot deduce the existence that easily.

Now assume that d fulfills conditions (A.3). For fixed $\alpha \in \mathcal{SR}$ this in particular implies

$$\Pi_T(\alpha + \lambda d) = \Pi_T(\alpha) + \lambda \Pi_T(d) \geq C \geq 0 \text{ (a.s.) } \forall \lambda \geq 0.$$

As $|\Pi_T(\alpha)| < \infty$ (a.s.), this inequality can only hold for all $\lambda \geq 0$ if $\Pi_T(d) \geq 0$ (a.s.). On the other hand $P(\Pi_T(d) > 0) > 0$ cannot hold either, because then d would be an arbitrage strategy due to $\Pi_0(d) = b^{\mathrm{T}}d = 0$. Thus $\Pi_T(d) = 0$ (a.s.).

Knowing this, it is easy to see that $d \in \mathbb{R}^m$ is a common direction of recession of Π_0 and $\mathcal{SR}$ if and only if

$$d \neq 0, \quad \Pi_0(d) = b^{\mathrm{T}}d = 0, \quad \Pi_T(d) = 0 \text{ (a.s.)}. \tag{A.4}$$

If no such vector d exists then, according to Theorem 1.10, the existence of a solution of the optimization problem is proven. However, if such a vector d exists, we will use the linear dependence as given by conditions (A.4) to reduce the dimension of the optimization problem.

Let $d = (d_1, \ldots, d_m)^{\mathrm{T}} \in \mathbb{R}^m$ be a vector satisfying (A.4) and assume without loss of generality that $d_m \neq 0$. Denote the unit vectors in $\mathbb{R}^m$ by $e^1, \ldots, e^m$. Then, according to (A.4)

$$\Pi_0(d) = \sum_{i=1}^m b_i d_i = 0 \ \Leftrightarrow\ b_m = -\sum_{i=1}^{m-1} \frac{d_i}{d_m} b_i$$
$$\Pi_T(d) = \sum_{i=1}^m d_i \Pi_T(e^i) = 0 \ \Leftrightarrow\ \Pi_T(e^m) = -\sum_{i=1}^{m-1} \frac{d_i}{d_m} \Pi_T(e^i),$$

where the latter equalities hold almost surely. This in turn implies that for arbitrary $\alpha \in \mathbb{R}^m$ the objective and constraint can be transformed to

$$\Pi_0(\alpha) = \sum_{i=1}^{m-1} b_i \alpha_i + \alpha_m \left(-\sum_{i=1}^{m-1} \frac{d_i}{d_m} b_i\right) = \sum_{i=1}^{m-1} \left(\alpha_i - \alpha_m \frac{d_i}{d_m}\right) b_i$$
$$\Pi_T(\alpha) = \sum_{i=1}^m \alpha_i \Pi_T(e^i) = \Pi_T\left(\sum_{i=1}^{m-1} \left(\alpha_i - \alpha_m \frac{d_i}{d_m}\right) e^i\right) \text{ (a.s.)}.$$

Thus, if we define $\tilde{\alpha}_i := \alpha_i - \alpha_m \frac{d_i}{d_m}$, $i = 1, \ldots, m-1$, optimization problem (A.1) is equivalent to the problem

$$\min_{\tilde{\alpha} \in \mathbb{R}^{m-1}} \Pi_0\begin{pmatrix} \tilde{\alpha} \\ 0 \end{pmatrix} \quad \text{s.t.} \quad \Pi_T\begin{pmatrix} \tilde{\alpha} \\ 0 \end{pmatrix} \geq C \text{ (a.s.)}. \tag{A.5}$$

In contrast to optimization problem (A.1), the equivalent problem (A.5) is reduced by one dimension. Of course the existence of a solution of problem (A.5) is equivalent to the existence of a solution of problem (A.1). The question is now whether the reduced problem has a solution.

Note that the analysis in the proof carried out so far can be repeated in analogy for the reduced problem (A.5). In particular the reduced problem has a solution if there is no common direction of recession of the objective and the feasible set, that is if there is no $\tilde{d} \in \mathbb{R}^{m-1}$ satisfying

$$\tilde{d} \neq 0, \quad \Pi_0 \begin{pmatrix} \tilde{d} \\ 0 \end{pmatrix} = (b_1, \ldots, b_{m-1})\tilde{d} = 0, \quad \Pi_T \begin{pmatrix} \tilde{d} \\ 0 \end{pmatrix} = 0 \text{ (a.s.)}.$$

In case there is a vector $\tilde{d}$ satisfying the above conditions, we can iteratively reduce the problem dimension until either there is no vector satisfying the conditions or until we have reduced the problem size to one variable $\tilde{\alpha} \in \mathbb{R}$. If the latter is the case, the optimization problem is given by

$$\begin{aligned} \min_{\tilde{\alpha} \in \mathbb{R}} \; & \Pi_0((\tilde{\alpha}, 0, \ldots, 0)^{\mathrm{T}}) = \tilde{\alpha} b_1 \\ \text{s.t. } \; & \Pi_T((\tilde{\alpha}, 0, \ldots, 0)^{\mathrm{T}}) = \tilde{\alpha} \Pi_T(e^1) \geq C \;\; \text{(a.s.)}. \end{aligned}$$

By assumption, the feasible set of the optimization problem is non-empty. Thus, as $C \geq 0$, $P(C > 0) > 0$, the random variable $\Pi_T(e^1)$ cannot be the zero vector. This in turn implies that $\Pi_T((\tilde{d}, 0, \ldots, 0)^{\mathrm{T}}) = \tilde{d}\Pi_T(e^1) = 0$, $\tilde{d} \in \mathbb{R}$, if and only if $\tilde{d} = 0$. Hence we can conclude that for the one-dimensional problem, the objective and the feasible set have no common directions of recession. Thus the one-dimensional problem has a solution and so does the original problem due to the equivalence of all reduced problems to the original problem. This concludes the proof. □

Note that the proof of the existence theorem solely depends on the primal problem, while analogous results in the case of dynamic trading strategies are also based on the associated dual problem, which is the maximization of an expected value with respect to a certain class of probability measures. However, in our case the primal problem is much easier, because far less strategies have to be considered. In fact the optimization problem is finite dimensional. In contrast to this the associated dual problem is more complicated as the class of probability measures within the dual problem is much larger. Nevertheless a more detailed analysis of the dual problem results in interesting interpretations from a financial point of view (see Section 4.2 for details).

Besides the existence of cost-optimal super-replicating strategies, another point of interest is the characterization of such strategies in case an exact hedging strategy exists in $\mathcal{S}$. As usual one can derive by simple arbitrage arguments that in such a case the cost-optimal super-replicating strategy is also an exact one. The result is stated in the following theorem.

Theorem A.5. *Suppose that Assumptions A.1 and A.2 hold and that Π_t and C are given as stated in Definition A.3. Further let the market model $\mathcal{M}$ be arbitrage-free in the set of trading strategies $\mathcal{S}$. If an exact hedging strategy exists in $\mathcal{S}$, i.e. $\Pi_T(\alpha) = C$ (a.s.) for some $\alpha \in \mathcal{S}$, every cost-optimal super-replicating $\mathcal{S}$-strategy is also exact and the prices at $t = 0$ coincide with the price of the exact hedge. In particular the cost-optimal super-replicating $\mathcal{S}$-strategy equals the exact hedge in $\mathcal{S}$, if the latter is unique.*

Proof. Let α_E denote an exact hedge in $\mathcal{S}$ and α_C a cost-optimal super-replicating $\mathcal{S}$-strategy. Due to $\Pi_T(\alpha_E) = C$ (a.s.), α_E is also a super-replicating $\mathcal{S}$-strategy. Thus $\Pi_0(\alpha_C) \leq \Pi_0(\alpha_E)$.

Assume that $\Pi_0(\alpha_C) < \Pi_0(\alpha_E)$ and let $\beta := \frac{\Pi_0(\alpha_E)}{\Pi_0(\alpha_C)}\alpha_C - \alpha_E$. Then $\Pi_0(\beta) = 0$ and

$$\Pi_T(\beta) = \frac{\Pi_0(\alpha_E)}{\Pi_0(\alpha_C)}\Pi_T(\alpha_C) - \Pi_T(\alpha_E) \geq C\left(\frac{\Pi_0(\alpha_E)}{\Pi_0(\alpha_C)} - 1\right) \geq 0$$

and greater than zero with positive probability. But this implies that β is an arbitrage strategy which contradicts the assumption. Thus $\Pi_0(\alpha_C) = \Pi_0(\alpha_E)$.

Now assume that the cost-optimal super-replicating $\mathcal{S}$-strategy is not exact, that is $\Pi_T(\alpha_C) > C$ with positive probability. Then $\gamma := \alpha_C - \alpha_E$ has zero cost and

$$\Pi_T(\gamma) = \Pi_T(\alpha_C) - \Pi_T(\alpha_E) \geq C - C = 0$$

and greater than zero with positive probability. This implies that γ is an arbitrage strategy which again contradicts the assumption.

Further, if the exact hedge in $\mathcal{S}$ is unique, it must coincide with the cost-optimal super-replicating $\mathcal{S}$-strategy, because the latter is itself an exact hedge. □

B Source Code

This appendix illustrates, based on Matlab 6.1 source code, how the algorithms presented in the previous chapters can be implemented in practice. Without loss of generality we will focus on Algorithm 3 to discuss the key steps of such an implementation. However, all semi-infinite optimization algorithms contained in this book including the extensions listed in Chapters 6 and 7 can be treated with slight modifications of the source code presented below.

In the following we aim at computing a robust static hedge portfolio based on Heston's stochastic volatility model driven by the stochastic differential equations

$$\begin{aligned} dS_t &= (r-\delta)S_t dt + \sqrt{Y_t} S_t dW_t^1 \\ dY_t &= \kappa(\theta - Y_t)dt + \xi\sqrt{Y_t} dW_t^2, \quad dW_t^1 dW_t^2 = \rho dt \\ dB_t &= rB_t dt. \end{aligned}$$

In this model a closed form solution for the value of standard calls is readily available (see Section 3.2 for a detailed description). We assume that this closed form solution has already been implemented in Matlab and can be called via

$$\text{HestonCall} = \text{HestonCall}(\kappa, \rho, \xi, \theta, \text{fwd}, \text{dfs}, K, T, S, Y_0), \tag{B.1}$$

where $\text{fwd} = \exp((r-\delta)T)$ and $\text{dfs} = \exp(-rT)$ denote the forward and discount factor from the current time until maturity T. Since this call pricing function will be evaluated hundreds or even thousands of times during the optimization, it is beneficial to code the closed form in efficient programming languages like C++ and to compile it as a Matlab MEX-file. This approach combines the advantages of the high level Matlab programming language with the speed of C++.

Before turning to the source code we briefly summarize the main features of the implementation. Algorithm 3 starts with an initially chosen hedge portfolio and a finite number of scenarios for which the super-replication property should hold. Since many other market scenarios in the parameter hypercube are not covered by this set, the hedge portfolio may lead to losses if the market differs from (or is not close to) a situation described by the finite number of scenarios. Consequently, we need to compute the worst case hedge error of the current portfolio, and add the scenario for which it occurs to the set of scenarios for which the super-replication property should hold. Then we can compute a new static superhedge based on this extended set of scenarios.

An implementation of this algorithm requires to numerically solve the static super-replication problem (5.15) subject to a finite set of scenarios as well as the constraint violation problem (5.16) determining the worst case hedging error. For both tasks we made use of Matlab solvers contained in the optimization toolbox. To be more precise, we used the solver linprog for the solution of the linear programming problems (5.15). The global nonlinear optimization problems (5.16) were tackled with a successive call of the solver fmincon with random start points. This cycle of linear and nonlinear optimization needs to be repeated until the worst case hedge error is smaller than a prespecified tolerance (in our case TOL $= 10^{-8}$).

The described iterative procedure is the heart of the source code presented in Listing B.1. As an example the problem parameters are set such that the algorithm computes the optimal hedge portfolio listed in Table 5.16, which corresponds to a diameter $\Delta = 15\%$ of the robustness intervals. As a side remark it should be mentioned that even if another programming language is chosen to implement the algorithm, the high level Matlab code listed below allows to quickly get an overview of the key steps.

Listing B.1: Main source code robusthedge.m

```
%=========================================
%==            robusthedge.m           ==
%=========================================

clear all;
warning off;

%global variables
global x S_0 r d T K D StandOptMaturities StandOptStrikes;

%-----------------------------------------
%- Problem and Optimization Parameters -
%-----------------------------------------

%optimization parameters
x = [1;0;0;0;0;0;0;0;0;0]; %initial hedge portfolio
x_lb = -50*ones(10,1);      %lower bounds on x (make the first unbounded problems bounded)
x_ub = 50*ones(10,1);       %upper bounds on x

TOL = 1e-8;                 %convergence tolerance: if slack >= -TOL, the algorithm stops
iter_max = 20;              %maximum number of main iterations of the algorithm
eps_semiinf = 1e-2;         %tolerance defining when slack minimizers will be added to the grid
eps_semiinf_factor = 10;    %adaptive factor reducing eps_semiinf in each iteration

num_slackmin = 50;          %number of random start points in each subdomain where the slack
        function is smooth (3 in this example)
rand('state',5);            %start seed for random points in global slack minimization
epsilon = 1e-6;             %defines termination criterion for slack minimizations
eps_bounds = 1e-3;          %defines when solutions of the slack minimizations are regarded as
        feasible

%market data
S_0 = 2750;
r = 0.055;
d = 0.025;

%Barrier option parameters
T = 1;
K = 2750;
D = 3300;

%Standard options to be considered in the hedge portfolio
StandOptMaturities = [1.0 1.0 1.0 1.0 1.0 0.75 0.75 0.75 0.5 0.5];
StandOptStrikes = [2750 3300 3350 3450 3600 3300 3400 3600 3300 3500];

%implied model parameters at t=0
v_0 = 0.04;           %start variance
kappa_0 = 1.5;        %mean reversion speed
theta_0 = 0.04;       %mean reversion level of variance
xi_0 = 0.2;           %volatility of variance
rho_0 = -0.5;         %correlation between spot and vol

%initial model parameter vector and parameter dimension
```

```
para_0 = [v_0;kappa_0;theta_0;xi_0;rho_0];
dim_para = length(para_0);

%definition of robustness intervals as centered multidimensional intervals
v_center = (0.2)^2;
kappa_center = 1.5;
theta_center = (0.2)^2;
xi_center = 0.2;
rho_center = -0.5;

v_radius = 0.075;
kappa_radius = 0.075;
theta_radius = 0.075;
xi_radius = 0.075;
rho_radius = 0.075;

para_lb = [max(0.001,(sqrt(v_center)-v_radius)^2);kappa_center-kappa_radius;max(0.001,(sqrt(
      theta_center)-theta_radius)^2);max(0.001,xi_center-xi_radius);rho_center-rho_radius]; %lower
      bounds for robustness interval
para_ub = [(sqrt(v_center)+v_radius)^2;kappa_center+kappa_radius;(sqrt(theta_center)+theta_radius)
      ^2;xi_center+xi_radius;rho_center+rho_radius]; %upper bounds for robustness interval

%-----------------
%- Initial output -
%-----------------

fprintf('\n');
fprintf('Starting optimization...\n');
fprintf('Tolerance for worst case hedge error: %4.2e\n',-TOL);
fprintf('\n');

%-------------------
%- Initialize grids -
%-------------------

N_t = 3;                 %initial number of grid points in time dimension
N_S = 100;               %initial number of grid points in stock dimension
para_N = [3;3;3;3;3];    %initial number of grid points in each parameter dimension

%initialize time grid (including option maturities)
deltat = T/N_t;
t = [StandOptMaturities 0:deltat:T];
t = sort(t);  %sort grid and eliminate same points in time
for j=1:length(t)
    for i=1:length(t)-1
        if t(i+1)==t(i)
            t(i+1)=T+1;
        end
    end
    t = sort(t);
end
for i=1:length(t)
    if t(i)==T
        N_t = i;
    end
end
t = t(1:N_t);

%initialize grid in each parameter dimension
for i=1:dim_para
    delta_para(i) = (para_ub(i)-para_lb(i))/para_N(i);
end

para1 = [para_lb(1):delta_para(1):para_ub(1)];
para_N(1) = length(para1);
para2 = [para_lb(2):delta_para(2):para_ub(2)];
para_N(2) = length(para2);
para3 = [para_lb(3):delta_para(3):para_ub(3)];
para_N(3) = length(para3);
para4 = [para_lb(4):delta_para(4):para_ub(4)];
para_N(4) = length(para4);
para5 = [para_lb(5):delta_para(5):para_ub(5)];
para_N(5) = length(para5);

%set up multi-dimensional time-parameter-grid (discretization of hypercube)
num_gridpoints = N_t; %compute total number of points of combined grid
for i=1:dim_para
    num_gridpoints = num_gridpoints*para_N(i);
end

t_para_grid = zeros(1,1+dim_para);
```

```
counter = 0;
for i=1:N_t
    for j=1:para_N(1)
        for k=1:para_N(2)
            for l=1:para_N(3)
                for m=1:para_N(4)
                    for n=1:para_N(5)
                        %only add those points satisfying the feller condition
                        if (para2(k)*para3(l)-0.5*para4(m)*para4(m)>=0)
                            t_para_grid(counter+1,1) = t(i);
                            t_para_grid(counter+1,2) = para1(j);
                            t_para_grid(counter+1,3) = para2(k);
                            t_para_grid(counter+1,4) = para3(l);
                            t_para_grid(counter+1,5) = para4(m);
                            t_para_grid(counter+1,6) = para5(n);
                            counter = counter+1;
                        end
                    end
                end
            end
        end
    end
end
num_gridpoints = counter;
clear counter;

%initialize stock-price grid for terminal time T
%Note that for the 10 considered options the interval [0,BarrierStrike] can be skipped
deltaS = (D-K)/N_S;
S = [K+deltaS:deltaS:D D];
if S(length(S))==S(length(S)-1) %eliminate D if contained twice
    S = S(1:length(S)-1);
end
N_S=length(S);

%-----------------------------------------------
%- %set up linear programming problem data -
%-          min c'x s.t. A*x<=b               -
%-----------------------------------------------

%set up objective function vector
c = zeros(1,length(StandOptMaturities));
for i=1:length(c)
    c(i) = HestonCall(para_0(2), para_0(5), para_0(4), para_0(3), exp((r-d)*StandOptMaturities(i)),
           exp(-r*StandOptMaturities(i)), StandOptStrikes(i), StandOptMaturities(i), S_0, para_0(1));
end

%define constraints: distinguish two cases of barrier hit and no barrier hit
%Case 1: Barrier is hit
A1 = zeros(num_gridpoints,length(c));
b1 = zeros(num_gridpoints,1);
for i=1:num_gridpoints
    for k=1:length(c)
        if StandOptMaturities(k)>t_para_grid(i,1)  %other options have previously expired worthless
            A1(i,k) = HestonCall(t_para_grid(i,3), t_para_grid(i,6), t_para_grid(i,5), t_para_grid(i
                ,4), exp((r-d)*(StandOptMaturities(k)-t_para_grid(i,1))), exp(-r*(
                StandOptMaturities(k)-t_para_grid(i,1))), StandOptStrikes(k), StandOptMaturities(k)
                -t_para_grid(i,1), D, t_para_grid(i,2));
        end
        if StandOptMaturities(k)==t_para_grid(i,1) %in case t=T set final payoff of option
            A1(i,k) = max(D-StandOptStrikes(k),0); %remark: if barrier is hit we have S_tau = D
        end
    end
end

%Case 2: Barrier is not hit
A2 = zeros(N_S,length(c));
b2 = zeros(N_S,1);
for i=1:N_S
    for j=1:length(c)
        if StandOptMaturities(j)==T
            A2(i,j) = max(S(i)-StandOptStrikes(j),0);
        end
    end
    b2(i) = max(S(i)-K,0); %barrier option payoff
end

%add box constraints on x to A2 and b2
for i=0:length(c)-1
    A2(N_S+2*i+1,i+1) = 1;
    A2(N_S+2*i+2,i+1) = -1;
```

```
    b2(N_S+2*i+1) = x_lb(i+1);
    b2(N_S+2*i+2) = -x_ub(i+1);
end
N_S = N_S+2*length(c);

%combine matrices for both cases in one constraint matrix
A = [-A1;-A2];
b = [-b1;-b2];
N = length(A1(:,1)); %store N since A1 will be dynamically adjusted

%-----------------------------------------------
%- Compute initial iterate and hedge error -
%-----------------------------------------------

iter = 0;

%Compute cheapest superhedge for initial grid
[x,fval,exitflag,output,lambda] = linprog(c,A,b,[],[],[],[],x,optimset('Display','off','LargeScale',
      'off'));
fval = fval/S_0;  %Transform cost of portfolio to basis points

%Start slack minimization
%Run a local nonlinear optimization for random points in the parameter set
i = 1;
hedge_error_worstcase = 0;
while (i<=3*num_slackmin)

    %choose random starting point for optimization
    %distinguish 3 regions of non-differentiability, i.e. [0,0.5], [0.5,0.75], [0.75,1.0]
    if (i<=3*num_slackmin) %region [0,0.5]
        rand_start = [0.5*rand(1);para_lb(1)+rand(1)*(para_ub(1)-para_lb(1));para_lb(2)+rand(1)*(
              para_ub(2)-para_lb(2));para_lb(3)+rand(1)*(para_ub(3)-para_lb(3));para_lb(4)+rand(1)*(
              para_ub(4)-para_lb(4));para_lb(5)+rand(1)*(para_ub(5)-para_lb(5))];
        t_lb = 0;
        t_ub = 0.5;
    end
    if (i<=2*num_slackmin) %region [0.5,0.75]
        rand_start = [0.5+(0.75-0.5)*rand(1);para_lb(1)+rand(1)*(para_ub(1)-para_lb(1));para_lb(2)+
              rand(1)*(para_ub(2)-para_lb(2));para_lb(3)+rand(1)*(para_ub(3)-para_lb(3));para_lb(4)+
              rand(1)*(para_ub(4)-para_lb(4));para_lb(5)+rand(1)*(para_ub(5)-para_lb(5))];
        t_lb = 0.5;
        t_ub = 0.75;
    end
    if (i<=1*num_slackmin) %region [0.75,1.0]
        rand_start = [0.75+(1.0-0.75)*rand(1);para_lb(1)+rand(1)*(para_ub(1)-para_lb(1));para_lb(2)
              +rand(1)*(para_ub(2)-para_lb(2));para_lb(3)+rand(1)*(para_ub(3)-para_lb(3));para_lb(4)
              +rand(1)*(para_ub(4)-para_lb(4));para_lb(5)+rand(1)*(para_ub(5)-para_lb(5))];
        t_lb = 0.75;
        t_ub = 1.0;
    end

    %minimize slack
    [tpara_opt,tpara_fvalue] = fmincon(@fnewton,rand_start,[],[],[],[],[t_lb;para_lb],[t_ub;para_ub
          ],@nonlcon,optimset('LargeScale','off','Display','off','TolX',epsilon,'GradConstr','on'));

    %check solution for feasibility (to not include results of unsuccessful optimizations)
    sol_feasible = 1;
    if (tpara_fvalue/S_0 < -TOL) %only include points if they violate the constraints by TOL
        if (tpara_opt(1)<t_lb-eps_bounds)|(tpara_opt(1)>t_ub+eps_bounds)
           sol_feasible = 0;
        end
        for j=1:dim_para
            if (tpara_opt(j+1)<para_lb(j)-eps_bounds)|(tpara_opt(j+1)>para_ub(j)+eps_bounds)
                sol_feasible = 0;
            end
        end
    end

    %add feasible points with negative slack to the grid A1 (for cases of barrier hit)
    if (sol_feasible==1)&(tpara_fvalue/S_0 < -eps_semiinf)
        if tpara_fvalue < hedge_error_worstcase
            hedge_error_worstcase = tpara_fvalue;

            index = length(t_para_grid(:,1));
            t_para_grid(index+1,:) = tpara_opt';
            A1(index+1,:)=zeros(1,length(c));
            b1(index+1) = 0;
            for k=1:length(c)
                if StandOptMaturities(k)>t_para_grid(index+1,1)
                    A1(index+1,k) = HestonCall(t_para_grid(index+1,3), t_para_grid(index+1,6),
                         t_para_grid(index+1,5), t_para_grid(index+1,4), exp((r-d)*(
```

```
                                    StandOptMaturities(k)-t_para_grid(index+1,1))), exp(-r*(StandOptMaturities(k)-t_para_grid(index+1,1))), StandOptStrikes(k), StandOptMaturities(k)-t_para_grid(index+1,1), D, t_para_grid(index+1,2));
                    end
                    if StandOptMaturities(k)==t_para_grid(index+1,1) %in case t=T set final payoff of option
                        A1(index+1,k) = max(D-StandOptStrikes(k),0);
                    end
                end
            end
        end
        i = i+1;
end
hedge_error_worstcase = hedge_error_worstcase/S_0;

%reformat arrays for LP-solver
A = [-A1;-A2];
b = [-b1;-b2];

%output of initial iterate
fprintf(' Iter.   #Points  #SlackMin      Cost               wc-error          epsilon             x(1)               x(2)               x(3)               x(4)               x(5)               x(6)               x(7)               x(8)               x(9)               x(10)\n');
fprintf(' %3g       %3g       %3g       %8.6e    %8.6e    %8.6e    %8.6e    %8.6e    %8.6e    %8.6e    %8.6e    %8.6e    %8.6e    %8.6e    %8.6e    %8.6e\n',iter,N+length(b2),i-1,fval,hedge_error_worstcase,eps_semiinf,x(1),x(2),x(3),x(4),x(5),x(6),x(7),x(8),x(9),x(10));

%reset arrays size and eps_semiinf tolerance
N = length(A1(:,1)); %store N since A1 is dynamically adjusted
eps_semiinf = eps_semiinf/eps_semiinf_factor;

%clear function memory (many calls of fnewton by the optimizer can exceed machine memory)
clear functions;

%--------------------------------------
%- Begin main optimization loop -
%--------------------------------------

hedge_error_worstcase = -1; %enter the loop at least once

%run at least 10 iterations to avoid early termination due to bad start points for
%global slack minimizations
while (iter<=iter_max)&(hedge_error_worstcase<-TOL)|(iter<=10)

    %Compute cheapest superhedge for current grid
    [x,fval,exitflag,output,lambda] = linprog(c,A,b,[],[],[],[],x,optimset('Display','off','LargeScale','off'));
    fval = fval/S_0;  %Transform cost of portfolio to basis points

    %Do a local nonlinear optimization for random points in the parameter set
    i = 1;
    hedge_error_worstcase = 0;
    while (i<=3*num_slackmin)
        %choose random starting point for optimization
        %distinguish 3 regions of non-differentiability, i.e. [0,0.5], [0.5,0.75], [0.75,1.0]
        if (i<=3*num_slackmin) %region [0,0.5]
            rand_start = [0.5*rand(1);para_lb(1)+rand(1)*(para_ub(1)-para_lb(1));para_lb(2)+rand(1)*(para_ub(2)-para_lb(2));para_lb(3)+rand(1)*(para_ub(3)-para_lb(3));para_lb(4)+rand(1)*(para_ub(4)-para_lb(4));para_lb(5)+rand(1)*(para_ub(5)-para_lb(5))];
            t_lb = 0;
            t_ub = 0.5;
        end
        if (i<=2*num_slackmin) %region [0.5,0.75]
            rand_start = [0.5+(0.75-0.5)*rand(1);para_lb(1)+rand(1)*(para_ub(1)-para_lb(1));para_lb(2)+rand(1)*(para_ub(2)-para_lb(2));para_lb(3)+rand(1)*(para_ub(3)-para_lb(3));para_lb(4)+rand(1)*(para_ub(4)-para_lb(4));para_lb(5)+rand(1)*(para_ub(5)-para_lb(5))];
            t_lb = 0.5;
            t_ub = 0.75;
        end
        if (i<=1*num_slackmin) %region [0.75,1.0]
            rand_start = [0.75+(1.0-0.75)*rand(1);para_lb(1)+rand(1)*(para_ub(1)-para_lb(1));para_lb(2)+rand(1)*(para_ub(2)-para_lb(2));para_lb(3)+rand(1)*(para_ub(3)-para_lb(3));para_lb(4)+rand(1)*(para_ub(4)-para_lb(4));para_lb(5)+rand(1)*(para_ub(5)-para_lb(5))];
            t_lb = 0.75;
            t_ub = 1.0;
        end

        %minimize slack
        [tpara_opt,tpara_fvalue] = fmincon(@fnewton,rand_start,[],[],[],[],[t_lb;para_lb],[t_ub;
```

```
                para_ub],@nonlcon,optimset('LargeScale','off','Display','off','TolX',epsilon,'
                GradConstr','on'));

        %check solution for feasibility (to not include results of unsuccessful optimizations)
        sol_feasible = 1;
        if (tpara_fvalue/S_0 < -TOL) %only include points if they violate the constraints by TOL
            if (tpara_opt(1)<t_lb-eps_bounds)|(tpara_opt(1)>t_ub+eps_bounds)
               sol_feasible = 0
            end
            for j=1:dim_para
                if (tpara_opt(j+1)<para_lb(j)-eps_bounds)|(tpara_opt(j+1)>para_ub(j)+eps_bounds)
                    sol_feasible = 0
                end
            end
        end

        %if solution is feasible and has negative slack, add it to the grid
        if (sol_feasible==1)&(tpara_fvalue/S_0 < -eps_semiinf)
            if tpara_fvalue < hedge_error_worstcase
                hedge_error_worstcase = tpara_fvalue;

                index = length(t_para_grid(:,1));
                t_para_grid(index+1,:) = tpara_opt';
                A1(index+1,:)=zeros(1,length(c));
                b1(index+1) = 0;
                for k=1:length(c)
                    if StandOptMaturities(k)>t_para_grid(index+1,1)
                        A1(index+1,k) = HestonCall(t_para_grid(index+1,3), t_para_grid(index+1,6),
                              t_para_grid(index+1,5), t_para_grid(index+1,4), exp((r-d)*(
                              StandOptMaturities(k)-t_para_grid(index+1,1))), exp(-r*(
                              StandOptMaturities(k)-t_para_grid(index+1,1))), StandOptStrikes(k),
                              StandOptMaturities(k)-t_para_grid(index+1,1), D, t_para_grid(index
                              +1,2));
                    end
                    if StandOptMaturities(k)==t_para_grid(index+1,1) %in case t=T set final payoff
                         of option
                        A1(index+1,k) = max(D-StandOptStrikes(k),0);
                    end
                end
            end
        end
        i = i+1;
    end
    hedge_error_worstcase = hedge_error_worstcase/S_0;
    iter = iter+1;

    %Output of current iterate
    fprintf(' %3g      %3g      %3g      %8.6e   %8.6e   %8.6e   %8.6e   %8.6e   %8.6e   %8.6e   %8
         .6e   %8.6e   %8.6e   %8.6e   %8.6e   %8.6e\n',iter,N+length(b2),i-1,fval,
         hedge_error_worstcase,eps_semiinf,x(1),x(2),x(3),x(4),x(5),x(6),x(7),x(8),x(9),x(10));

    %reformat arrays for LP-solver
    A = [-A1;-A2];
    b = [-b1;-b2];
    N = length(A1(:,1)); %store N since A1 is dynamically adjusted

    %Update of optimization parameters
    eps_semiinf = eps_semiinf/eps_semiinf_factor;
    if (eps_semiinf<TOL)
        eps_semiinf = TOL;
    end

    %clear function memory (many calls of fnewton by the optimizer can exceed machine memory)
    clear functions;
end

%-----------------
%- Final output -
%-----------------

fprintf('\n');
if (hedge_error_worstcase>=-TOL)
    fprintf('Optimization terminated successfully.\n');
    fprintf('Worst case hedge error < -TOL.\n');
else
    fprintf('Maximum numbers of iterations reached.\n');
    fprintf('Worst case hedge error not yet < -TOL.\n');
end
fprintf('\n');
```

Lines 1–79 of Listing B.1 define the problem parameters as well as the main algorithm settings. In Lines 80–164 we construct the initial set of market scenarios for which the super-replication property should hold. The set of points t_para_grid reflects the set M_1 in Algorithm 3. The vector S corresponds to the set M_2. Since the value of the hedge portfolio at maturity of the barrier option is a piecewise linear function, it is sufficient to include the strikes of the calls with the same maturity and the barrier itself in the set M_2. However, here we chose an equidistant discretization of the one-dimensional interval $[K, D]$.

Afterwards lines 165–216 set up the problem data for the linear programming problem (5.15). In this process the two cases of a barrier hit and no barrier hit are distinguished. If a market scenario requires the evaluation of a call option before its maturity, we do so by calling the Heston closed form (B.1). In lines 217–226 we use the generated matrices and vectors to solve problem (5.15) with the Matlab linear programming algorithm linprog.

The computed hedge portfolio is then tested in lines 227–289 for a violation of the super-replication constraints. To determine the worst case hedge error, that is to solve the slack minimization (5.16), we split the domain into three parts divided by the lines of non-differentiability of the hedge error function in time direction. Since these lines of non-differentiability are given by the maturities of the calls which expire before the maturity of the barrier option, we divide the time domain into the three parts $[0, 0.5]$, $[0.5, 0.75]$ and $[0.75, 1.0]$. On the resulting subsets of the parameter hypercube the hedge error function (see Listing B.2) is smooth and can be minimized with Matlab's nonlinear programming solver fmincon (see line 252). As we are only interested in those market scenarios that satisfy the Feller condition, we include this constraint as the Matlab function nonlcon (see Listing B.3 for the source code) in the optimization process. To reduce the risk of getting stuck in a local minimum of problem (5.16) we repeat the minimization on each of the three subsets 50 times with random start point within the respective set. The computed worst case hedge error is then stored in the variable hedge_error_worstcase.

Listing B.2: Hedge error function fnewton.m

```
%====================================
%==            fnewton.m           ==
%====================================

function y = fnewton(z)

%global variables
global x S_0 r d T K D StandOptMaturities StandOptStrikes;

%compute value of hedge portfolio at point z
y = 0;
for k=1:10
    if StandOptMaturities(k)>z(1)
        y = y + x(k)*HestonCall(z(3), z(6), z(5), z(4), exp((r—d)*(StandOptMaturities(k)—z(1))), exp
             (—r*(StandOptMaturities(k)—z(1))), StandOptStrikes(k), StandOptMaturities(k)—z(1), D, z
             (2));
    end
end
```

Listing B.3: Feller constraint nonlcon.m

```
%===================================
%==          nonlcon.m            ==
%===================================

function [c,ceq,GC,GCeq] = nonlcon(z)

c = 0.5*z(5)*z(5)-z(3)*z(4);        %Feller constraint
ceq = [];                            %No equality constraint

GC = [0;0;-z(4);-z(3);z(5);0];  %Gradient of Feller constraint
GCeq = [];
```

After completing this initial step lines 290–304 update some of the parameters and write the computed hedge portfolio as well as the most relevant algorithm status variables to the command prompt. The update of the parameters also includes a reduction of the parameter ϵ_k in Algorithm 3. Here we chose the simple sequence eps_semiinf = eps_semiinf / eps_semiinf_factor, with eps_semiinf_factor = 10.

The cycle of solving the discrete super-replication problem (5.15) and computing the worst case constraint violation (5.16) is then repeated successively in the main optimization loop described in lines 305–398. This process is terminated if either the worst case hedge error is smaller than −TOL or if the number of iterations exceeds iter_max = 20 (see line 313). Finally, lines 399–411 inform the user about the optimization result with a brief output.

List of Figures

List of Tables

List of Algorithms

List of Code Listings

Bibliography

[1] F. Alizadeh and D. Goldfarb, *Second-order cone programming*, Math. Program. 95 (2003), pp. 3–51. ISMP 2000, Part 3 (Atlanta, GA).

[2] S. Allen and O. Padovani, *Risk Management Using Quasi-static Hedging*, Economic Notes 31 (2002), pp. 277–336.

[3] L. Andersen and R. Brotherton-Ratcliffe, *Exact Exotics*, Risk 9 (1996), pp. 85–89.

[4] M. Avellaneda, A. Levy, and A. Paras, *Pricing and Hedging Derivative Securities in Markets with Uncertain Volatilities*, Applied Mathematical Finance 2 (1995), pp. 73–88.

[5] G Bakshi, C. Cao, and Z. Chen, *Empirical Performance of Alternative Option Pricing Models*, Journal of Finance 52 (1997), pp. 2003–2049.

[6] D. Bates, *Jumps and Stochastic Volatility: Exchange Rate Processes Implicit in Deutsche Mark Options*, The Review of Financial Studies 9 (1996), pp. 69–107.

[7] H. Bauer, *Measure and integration theory*, de Gruyter Studies in Mathematics 26. Walter de Gruyter & Co., Berlin, 2001.

[8] R. Bellman, *Adaptive control processes: A guided tour*. Princeton University Press, Princeton, N.J., 1961.

[9] A. Ben-Tal and A. Nemirovski, *Lectures on Modern Convex Optimization: Analysis, Algorithms, and Engineering Applications*, MPS/SIAM Series on Optimization. Society for Industrial and Applied Mathematics (SIAM), 2001.

[10] S. J. Benson, Y. Ye, and X. Zhang, *Solving Large-Scale Sparse Semidefinite Programs for Combinatorial Optimization*, SIAM Journal on Optimization 10 (1999), pp. 443–461.

[11] L. Bergomi, *Smile Dynamics*, Risk Magazine (2004), pp. 117–123.

[12] J. F. Bonnans, J. C. Gilbert, C. Lemaréchal, and C. A. Sagastizábal, *Numerical optimization*, Universitext. Springer-Verlag, Berlin, 2003, Theoretical and practical aspects.

[13] J. F. Bonnans and A. Shapiro, *Perturbation Analysis of Optimization Problems*. Springer, 2000.

[14] Deutsche Börse, *Guide to the Equity Indices of Deutsche Börse*, 5.12. ed.. Deutsche Börse, 60485 Frankfurt am Main, Germany, 2006.

[15] I. Bouchouev and V. Isakov, *Uniqueness, stability and numerical methods for the inverse problem that arises in financial markets*, Inverse Problems 15 (1999), pp. R95–R116.

[16] J. Bowie and P. Carr, *Static Simplicity*, Risk 7 (1994), pp. 45–49.

[17] K. A. Boyle, T. F. Coleman, and Y. Li, *Hedging a Portfolio of Derivatives by Modeling Cost*. IEEE Proceedings of the 2003 International Conference on Computational Intelligence for Financial Engineering, 2003.

[18] M. Broadie and Ö. Kaya, *Exact Simulation of Stochastic Volatility and Other Affine Jump Diffusion Processes*, Operations Research 54 (2005).

[19] H. M. Brown, D. Hobson, and L. C. G. Rogers, *Robust Hedging of Barrier Options*, Mathematical Finance 11 (2001), pp. 285–314.

[20] H. Buehler, *Stochastic Volatility Models and Products*, Presentation at the Risk Training Course, Hong Kong, July 7/8, 2004.

[21] J. V. Burke, A. S. Lewis, and M. L. Overton, *Two Numerical Methods for Optimizing Matrix Stability*, Linear Algebra and its Applications 351–352 (2002), pp. 117–145.

[22] J. V. Burke, A. S. Lewis, and M. L. Overton, *Optimization and Pseudospectra, with Applications to Robust Stability*, SIAM Journal on Matrix Analysis and Applications 25 (2003), pp. 80–104.

[23] P. Carr and A. Chou, *Breaking Barriers*, Risk (1997), pp. 139–145.

[24] P. Carr, K. Ellis, and V. Gupta, *Static Hedging of Exotic Options*, Journal of Finance 53 (1998), pp. 1165–1190.

[25] T. F. Coleman, Y. Li, and M. Patron, *Discrete Hedging under Piecewise Linear Risk Minimization*, Journal of Risk 5 (2003), pp. 39–65.

[26] J. C. Cox, J. E. Ingersoll, and S. A. Ross, *A Theory of the Term Structure of Interest Rates.*, Econometrica 53 (1985), pp. 385–407.

[27] E. Derman, D. Ergener, and I. Kani, *Static options replication*, The Journal of Derivatives 2 (1995), pp. 78–95.

[28] B. Dumas, J. Fleming, and R. E. Whaley, *Implied Volatility Functions: Empirical Tests*, Journal of Finance 53 (1998), pp. 2059–2106.

[29] D. Y. Dupont, *Hedging Barrier Options: Current Methods and Alternatives*, EURANDOM - TUE, Report, 2001.

[30] N. El Karoui and M. C. Quenez, *Dynamic Programming and Pricing of Contingent Claims in an Incomplete Market*, SIAM Journal on Control and Optimization 33 (1995), pp. 29–66.

[31] R. J. Elliott and P. E. Kopp, *Mathematics of financial markets*, second. ed., Springer Finance. Springer-Verlag, New York, 2005.

[32] B. Engelmann, M. R. Fengler, M. Nalholm, and P. Schwendner, *Static versus Dynamic Hedges: An Empirical Comparison for Barrier Options*, Review of Derivatives Research 9 (2007), pp. 239–264.

[33] B. Engelmann, M. R. Fengler, and P. Schwendner, *Better than its Reputation: An Empirical Hedging Analysis of the Local Volatility Model for Barrier Options*, Available at SSRN: http://ssrn.com/abstract=935951, October 2006.

[34] B. Eraker, *Do Stock Prices and Volatility Jump? Reconciling Evidence from Spot and Option Prices*, Journal of Finance 59 (2004), pp. 1367–1403.

[35] B. Eraker, M. Johannes, and N. Polson, *The Impact of Jumps in Volatility and Returns*, Journal of Finance 63 (2003), pp. 1269–1300.

[36] B. Fares, D. Noll, and P. Apkarian, *Robust control via sequential semidefinite programming*, SIAM Journal on Control and Optimization 40 (2002), pp. 1791–1820.

[37] M. R. Fengler, *Arbitrage-free smoothing of the implied volatility surface*, SfB 649, Discussion paper, Humboldt-Universität zu Berlin, 2005.

[38] J. Fink, *An Examination of the Effectiveness of Static Hedging in the Presence of Stochastic Volatility*, The Journal of Futures Markets 23 (2003), pp. 859–890.

[39] R. Fletcher, *Practical Methods of Optimization*, 2nd. ed.. Wiley, May 2000.

[40] H. Föllmer and M. Schweizer, *Hedging by Sequential Regression: An Introduction to the Mathematics of Option Trading*, The ASTIN Bulletin 18 (1989), pp. 147–160.

[41] J.-P. Fouque, G. Papanicolaou, and K. R. Sircar, *Derivatives in Financial Markets with Stochastic Volatility*. Cambridge University Press, 2000.

[42] A. Friedman, *Partial differential equations of parabolic type*. Prentice-Hall Inc., 1964.

[43] Jim Gatheral, *The Volatility Surface, A Practitioner's Approach*. John Wiley and Sons, Inc., 2006.

[44] F. Gerlich, A. Giese, J. Maruhn, and E. Sachs, *Parameter identification in stochastic volatility models with time-dependent model parameters*, University of Trier, Technical report, 2006.

[45] A. M. Giese and J. H. Maruhn, *Cost-Optimal Static Super-Replication of Barrier Options - An Optimization Approach*, Journal of Computational Finance 10 (2007), pp. 71–97.

[46] M. A. Goberna and M. A. López, *Linear Semi-Infinite Optimization*, Wiley Series in Mathematical Methods in Practice. Wiley, 1998.

[47] D. Goldfarb and G. Iyengar, *Robust portfolio selection problems*, Mathematics of Operations Research 28 (2003), pp. 1–38.

[48] M. Grant, S. Boyd, and Y. Ye, *cvx Users' Guide*, http://www.stanford.edu/~boyd/cvx, 2007.

[49] D. Henrion, *H-infinity controller design on the COMPLIB problems with the Robust Control Toolbox for Matlab*, LAAS-CNRS, Report no. 05581, October 2005.

[50] S. L. Heston, *A Closed-Form Solution for Options with Stochastic Volatility with Applications to Bond and Currency Options*, The Review of Financial Studies 6 (1993), pp. 327–343.

[51] R. Hettich and K. O. Kortanek, *Semi-infinite programming: theory, methods, and applications*, SIAM Review 35 (1993), pp. 380–429.

[52] R. Hettich and P. Zencke, *Numerische Methoden der Approximation und semi-infiniten Optimierung*. Teubner, 1982.

[53] R. Horn and C. Johnson, *Matrix Analysis*. Cambridge University Press, 1985.

[54] A. Irle, *Finanzmathematik*, second. ed., Teubner Mathematical Textbooks. B. G. Teubner, Wiesbaden, 2003, Die Bewertung von Derivaten.

[55] M. S. Joshi, *The concepts and practice of mathematical finance*, Mathematics, Finance and Risk. Cambridge University Press, Cambridge, 2003.

[56] C. Kanzow, C. Nagel, H. Kato, and M. Fukushima, *Successive Linearization Methods for Nonlinear Semidefinite Programs*, Computational Optimization and Applications 31 (2005), pp. 251–273.

[57] I. Karatzas and S. E. Shreve, *Brownian motion and stochastic calculus*, second. ed., Graduate Texts in Mathematics 113. Springer-Verlag, New York, 1991.

[58] I. Karatzas and S. E. Shreve, *Methods of Mathematical Finance*, Applications of Mathematics (New York) 39. Springer-Verlag, New York, 1998.

[59] P. E. Kloeden and E. Platen, *Numerical solution of stochastic differential equations*, Applications of Mathematics (New York) 23. Springer-Verlag, Berlin, 1999.

[60] P. E. Kloeden, E. Platen, and H. Schurz, *Numerical solution of SDE through computer experiments*, Universitext. Springer-Verlag, Berlin, 1994.

[61] M. Kočvara, F. Leibfritz, M. Stingl, and D. Henrion, *A nonlinear SDP algorithm for static output feedback problems in COMPl$_e$ib*. Proceedings of the 16th IFAC World Congress on Automatic Control, July 2005.

[62] F. Leibfritz, *A LMI–based algorithm for designing suboptimal static $\mathcal{H}_2/\mathcal{H}_\infty$ output feedback controllers*, SIAM Journal on Control and Optimization 39 (2001), pp. 1711–1735.

[63] F. Leibfritz, *Static output feedback $\mathcal{H}_\infty$–synthesis by solving the $\mathcal{H}_2/\mathcal{H}_\infty$–NSDP*. Proceedings in Applied Mathematics and Mechanics (PAMM), 5, December 2005.

[64] F. Leibfritz and J. Maruhn, *A Successive SDP-NSDP Approach to a Robust Optimization Problem in Finance*, (2008). To appear in Journal of Computational Optimization and Applications.

[65] F. Leibfritz and E. M. E. Mostafa, *An interior point constrained trust region method for a special class of nonlinear semi-definite programming problems*, SIAM Journal on Optimization 12 (2002), pp. 1048–1074.

[66] F. Leibfritz and S. Volkwein, *Numerical feedback controller design for PDE systems using model reduction: Techniques and case studies*, To appear in SIAM Book Series on Computational Science and Engineering, SIAM, 2005.

[67] J. H. Maruhn, *Duality in Static Hedging of Barrier Options*, Optimization 58 (2009), pp. 319–333.

[68] J. H. Maruhn, M. Nalholm, and M. R. Fengler, *Empirically Robust Static Uncertain Skew Hedges for Reverse Barrier Options*, To appear in Quantitative Finance (2009).

[69] J. H. Maruhn and E. W. Sachs, *Robust Static Super-Replication of Barrier Options in the Black Scholes Model*. Robust Optimization-Directed Design (A. J. Kurdila, P. M. Pardalos, and M. Zabarankin, eds.), Series on Nonconvex Optimization and its Applications, pp. 127–143. Springer, 2006.

[70] J. H. Maruhn and E. W. Sachs, *Robust Static Hedging of Barrier Options in Stochastic Volatility Models*, (2008). To appear in Mathematical Methods of Operations Research.

[71] S. Mizohata, *Unicité du prolongement des solutions pour quelques operatéurs differentiels paraboliques*, Mem. Coll. Sci. Univ. Kyoto Ser. A Math. 31 (1958), pp. 219–239.

[72] M. Nalholm and R. Poulsen, *Static Hedging and Model Risk for Barrier Options*, Journal of Futures Markets 26 (2006), pp. 449–463.

[73] M. Nalholm and R. Poulsen, *Static Hedging of Barrier Options under General Asset Dynamics: Unification and Application*, Journal of Derivatives 13 (2006), pp. 46–60.

[74] A. Neuberger and S. D. Hodges, *Rational Bounds on the Prices of Exotic Options*, London Business School, EFA Working Paper, Report no. 0280, July 2000.

[75] B. Øksendal, *Stochastic differential equations*, fifth. ed., Universitext. Springer-Verlag, Berlin, 2000, An introduction with applications.

[76] T. Pennanen and M. Koivu, *Epi-convergent discretizations of stochastic programs via integration quadratures*, Numerische Mathematik 100 (2005), pp. 141–163.

[77] R. T. Rockafellar, *Convex Analysis*, Princeton Mathematical Series, No. 28. Princeton University Press, Princeton, N.J., 1970. MR MR0274683 (43 #445)

[78] C. W. Scherer, *LMI relaxations in robust control*, European Journal of Control 12 (2006).

[79] U. Schmock, S. E. Shreve, and U. Wystup, *Valuation of exotic options under shortselling constraints*, Finance and Stochastics 6 (2002), pp. 143–172.

[80] A. Shapiro, *Monte Carlo Sampling Methods*, Stochastic Programming (A. Ruszczyński and A. Shapiro, eds.), Handbooks in Operations Research and Management 10, Elsevier Science B.V., 2003, pp. 353–425.

[81] A. N. Shiryaev, *Probability*, second. ed., Graduate Texts in Mathematics 95. Springer-Verlag, New York, 1996.

[82] J. F. Sturm, *Using SeDuMi 1.02, a*, Optimization Methods and Software 11 (1999), pp. 625–653.

[83] K. B. Toft and C. Xuan, *How Well Can Barrier Options Be Hedged by a Static Portfolio of Standard Options?*, The Journal of Financial Engineering 7 (1998), pp. 147–175.

[84] K. C. Toh, M. J. Todd, and R. H. Tütüncü, *SDPT3 – a MATLAB Software package for Semidefinite Programming, version 2.1*, Optimization Methods and Software 11 (1999), pp. 545–581.

[85] R. H. Tütüncü and M. Koenig, *Robust asset allocation*, Annals of Operations Research 132 (2004), pp. 157–187.

[86] L. Vandenberghe and S. Boyd, *Semidefinite programming*, SIAM Review 38 (1996), pp. 49–95.

[87] P. Wolfe, *A Duality Theorem for Nonlinear Programming*, Quarterly of Applied Mathematics 19 (1961), pp. 239–244.

[88] S. J. Wright, *Algorithms and Software for Linear and Nonlinear Programming*. Foundations of Computer-Aided Process Design. CACHE Publications, 1999.

[89] P. G. Zhang, *Exotic Options - A Guide to Second Generation Options*. World Scientific, 1998.

Symbols and Abbreviations

Abbreviations

a.s.	Almost surely
ATM	At-the-money
BC	Bowie and Carr [16]
bp(s)	Basis point(s): One basis point equals 0.01%
BS	Black-Scholes
CV	Constraint violation
DEK	Derman, Ergener and Kani [27]
DOP	Down-and-out put
DV	Delta-vega
HSV	Heston stochastic volatility
LV	Local volatility
MA	Minimal assumptions
NLP	Nonlinear program
NSDP	Nonlinear semidefinite program
OTC	Over the counter: Financial instruments which are traded through a dealer network instead of a centralized exchange
SDP	Semidefinite program
SOCP	Second order cone program
STR	Strike spread
UOC	Up-and-out call

Basic Operations

$(a)^+$	Maximum of $a \in \mathbb{R}$ and zero, i.e. $(a)^+ = \max\{a, 0\}$
$a \wedge b$	Minimum of the reals a and b

Vectors, Matrices and Operations

$\|\cdot\|_2$	Euclidean vector norm defined by $\|x\|_2 = (\sum_i x_i^2)^{1/2}$
$\|\|\cdot\|\|_2$	Matrix lub-norm defined by $\|\|A\|\|_2 = \sup_{\|x\|_2=1} \|Ax\|_2$
$A \succ 0$	The real symmetric matrix A is positive definite
$A \succeq 0$	The matrix $A \in \mathbb{S}^m$ is positive semidefinite
$\lambda_{\min}(A)$	Minimal Eigenvalue of a matrix $A \in \mathbb{S}^m$
$\mathrm{Tr}(\cdot)$	Trace operator
$\langle A_1, A_2 \rangle$	Trace of $A_1^{\mathrm{T}} A_2$ for $A_1, A_2 \in \mathbb{R}^{l \times m}$
I_m	Identity matrix in $\mathbb{R}^{m \times m}$
$E(t, y, p)$	Matrix defining the ellipsoid around the model prices

Probabilistic Symbols

ω	Sample contained in the set Ω
Ω	Set of samples
$\mathcal{F}$	Sigma-algebra of the probability space $(\Omega, \mathcal{F}, P)$
$\mathcal{F}_t$	Element of the filtration $(\mathcal{F}_t)_{t \in I}$
$\mathcal{B}$	Borel sigma-algebra on $\mathbb{R}$
P	Probability measure of the probability space $(\Omega, \mathcal{F}, P)$
Q	Equivalent martingale measure
$Q^{(X,Y)}$	Joint distribution of the random variables X, Y under Q
E_Q	Expectation with respect to the measure Q
$E_Q(\cdot \mid \mathcal{F}_t)$	Conditional expectation relative to the sigma algebra $\mathcal{F}_t$
$\text{supp}(\mu)$	Support of the measure μ
$\mu\vert_B$	Measure μ restricted to the set B, $\mu\vert_B(\cdot) = \mu(\cdot \cap B)/\mu(B)$
$\Phi(\cdot)$	Cumulative normal distribution
W_t	Brownian motion / Wiener process at time t

Hedge Portfolio and Target Option

n	Number of standard options C^i in the hedge portfolio
α_i	Portfolio positions for the calls C^i and the bond B
$\alpha_i^{\text{lb}}, \alpha_i^{\text{ub}}$	Lower and upper bounds on the portfolio positions
C^i	Standard call i in the hedge portfolio
C_t^i	Value of call i at time t
C_{KO}	Payoff of a knock-out barrier option
C_{uo}	Payoff of an up-and-out call
T_i	Maturity of call C^i
T	Maturity of the barrier option
K_i	Strike of standard call C^i
K	Strike of the barrier option
D	Barrier of the barrier option
τ	Stopping time of the first barrier event
Π_t	Value of the hedge portfolio at time t
ϕ	Stochastic portfolio trading strategy $(\phi_t)_{t \in I}$

Financial Market Model

$\mathcal{M}$	Financial market model
B_t	Value of the bond at time t
S_t	Stock price at time t
Y_t	Process driving the stochastic volatility at time t, e.g. the variance
r	Riskfree rate
δ	Dividend yield
κ	Mean reversion speed of the variance

θ	Long-run mean of the variance
ξ	Volatility of volatility
ρ	Correlation of Brownian motions in stochastic volatility models
σ	Volatility
σ_0	Initial volatility at time $t = 0$
$\sigma_{\min}$	Lower bound of the volatility uncertainty interval
$\sigma_{\max}$	Upper bound of the volatility uncertainty interval
t	Time parameter $t \in I$
y	State of $Y_t(\omega) \in \mathcal{Y}$
$Y_{\min}, Y_{\max}$	Lower and upper bounds on y
p	Model parameters of the stochastic volatility model

Sets and Set Operations

A^c	Complement of the set A
$\bar{A}$	Closure of A
$\|A\|$	Cardinality of A
$L^1(P)$	Space of measurable functions whose absolute value has a finite integral with respect to P
$\mathbb{R}_+$	Set of non-negative real numbers $[0, \infty)$
$\mathbb{S}^m$	Space of real symmetric $m \times m$ matrices
$\mathcal{C}$	Set of standard calls $C^1, \dots, C^n$
I	Time index set
$\mathcal{Y}$	Interval with $Y_t \in \mathcal{Y} \; \forall t$ (a.s.)
SF	Set of self-financing trading strategies
$\mathcal{S}_{\mathrm{KO}}$	Set of static knock-out trading strategies
$\mathcal{S}\mathcal{R}_{\mathrm{KO}}$	Static super-replicating knock-out trading strategies, also denoted by $\mathcal{S}\mathcal{R}$. $\mathcal{S}\mathcal{R}(\sigma_{\min}, \sigma_{\max})$ contains those superhedges which are robust against movements of the volatility parameter $\sigma \in [\sigma_{\min}, \sigma_{\max}]$
Θ_1	Set of combinations (t, Y_t) for which the barrier might be hit
Θ_2	Stock prices S_T attainable without a barrier hit
Γ_1	Subset of $\Theta_1 \times U$
Γ_2	Equals the set $[0, D]$
L	L-shaped set $L = [0, T] \times \{D\} \cup \{T\} \times [0, D]$
U	Uncertainty set for the model parameters p
Δ	Size of the model parameter uncertainty set U

Optimization and Discretization Parameters

k	Iteration index
M	Number of samples for the Monte Carlo simulation
M_1, M_2	Discretizations of the parameter sets $\Gamma_1 \subset \Theta_1 \times U$ and Θ_2

Index

Bei Fragen zur Produktsicherheit wenden Sie sich bitte an:
If you have any questions regarding product safety,
please contact:

Walter de Gruyter GmbH
Genthiner Straße 13
10785 Berlin
productsafety@degruyterbrill.com